MEP 016B 3KW Generator Depot Maintenance Manual TM 5-6115-615-24P

Generator Set Skid Mounted, Tactical Quiet

edited by
Brian Greul

The MEP series of Military Generators are rugged, durable and incorporate proven diesel engine technology. This book is the depot maintenance manual. It also includes the repair parts and special tools list. It is being republished to assist enthusiasts, restorers, and aftermarket owners who use or wish to use these generators outside of military use.

An 8.5x11 3 hole punched loose leaf copy may be purchased for your 3 ring binder. Email books@ocotillopress.com for current information.

Should you have suggestions or feedback on ways to improve this book please send email to Books@OcotilloPress.com

---

Edited 2021 Ocotillo Press
ISBN 978-1-954285-26-2

Printed in the United States of America

Ocotillo Press
Houston, TX 77017
Books@OcotilloPress.com

MARINE CORPS STOCK LIST SL-4-05926B/06509B-24P/2
ARMY TECHNICAL MANUAL TM 5-6115-615-24P
NAVY PUBLICATION NAVFAC P-8-646-24P
AIR FORCE TECHNICAL ORDER TO 35C2-3-386-34

TECHNICAL MANUAL

ORGANIZATIONAL, INTERMEDIATE (FIELD), (DIRECT AND GENERAL SUPPORT), AND DEPOT MAINTENANCE REPAIR PARTS AND SPECIAL TOOLS LIST

GENERATOR SET, DIESEL ENGINE DRIVEN, TACTICAL, SKID MOUNTED, 3 KW, 3 PHASE 120/208 AND SINGLE PHASE 120/240 VOLTS AC AND 28 VOLTS DC

DOD MODEL	CLASS	MODE	NSN
MEP-016B	UTILITY	60 HZ	6115-01-150-4140
MEP-021B	UTILITY	400 HZ	6115-01-151-8126
MEP-026B	UTILITY	28 VDC	6115-01-150-0367

THIS PUBLICATION IS REQUIRED FOR OFFICIAL USE FOR ADMINISTRATIVE OR OPERATIONAL PURPOSES ONLY. DISTRIBUTION IS LIMITED TO U.S. GOVERNMENT AGENCIES. OTHER REQUESTS FOR THIS DOCUMENT MUST BE REFERRED TO COMMANDANT OF THE MARINE CORPS (HQSP-2) WASHINGTON, D.C. 20380-0001.

DESTRUCTION NOTICE FOR UNCLASSIFIED, LIMITED DOCUMENTS, DESTROY BY ANY METHOD THAT WILL PREVENT DISCLOSURE OF CONTENTS OR RECONSTRUCTION OF THE DOCUMENT.

PUBLISHED UNDER THE AUTHORITY OF HEADQUARTERS U.S. MARINE CORPS, THE DEPARTMENTS OF THE ARMY, AIR FORCE, AND NAVY

DECEMBER 1987

PCN 124 059270 00

DEPARTMENT OF THE NAVY
Headquarters, U.S. Marine Corps
Washington, DC 20380-0001

SL-4-05926B/06509B-24P/2
Change 2
PCN 124 059270 02
2 November 1992

Encl: (1) Replacement Pages

1. Purpose. To transmit replacement pages to the basic Manual, SL-4-05926B/06509B-24P/2, dated December 1987, Generator Set, Diesel Engine Driven, Tactical, Skid Mounted, 3KW, 3 Phase 120/208 and Single Phase 120/240 Volts AC and 28 Volts DC, MEP-016B, NSN 6115-01-150-4140, MEP-021B, NSN 6115-01-151-8126, and MEP-026B, NSN 6115-01-150-0367.

2. Action. Remove present pages listed below and replace with corresponding pages contained in the enclosure. Significant changes contained in the replacement pages of this Change are denoted by a bar (x) symbol.

PAGE ACTION

A/(B blank) A/(B blank) 9 thru 18/(19 blank) 9 thru 18/(19 blank) 21 thru 36/(37 blank) 21 thru 36/(37 blank)
39 thru 44/(45 blank) 39 thru 44/(45 blank) 47 thru 50/(51 blank) 47 thru 50/(51 blank) 53 thru 56/(57 blank)
53 thru 56/(57 blank) 59 and 60/(61 blank) 59 and 60/(61 blank) 63 thru
72/(73 blank) 63 thru 72/(73 blank) 74 thru 106 74 thru 106 109 and 110/(111
blank) 109 and 110/(111 blank) 113 and 114/(115 blank) -113 and 114/(115 blank) 117 thru 124/(125 blank)
117 thru 124/(125 blank) 127 and 128 127 and 128 131 thru 134
131 thru 134 137 and 138/(139 blank) 137 and 138/(139 blank) 141 thru 148/(149 blank) 141 thru 148/
(149 blank) 151 and 152/(153 blank) 151 and 152/(153 blank) 155 thru 158/(159 blank) 155 thru 158/
(159 blank) 161 and 162/(163 blank) -161 and 162/(163 blank) 165/(166 blank) 165/(166 blank) 196
thru 219/(220 blank) 196 thru 219/(220 blank)

SL-4-05926B/06509B-24P/2
Change 2

3. Filing Instructions. This Change transmittal page will be filed immediately following the signature page of the basic Manual.

BY DIRECTION OF THE COMMANDANT OF THE MARINE CORPS

OFFICIAL

FRED J. HOWARD
Director, Technical Support Division
Marine Corps Logistics Bases
Albany, Georgia 31704-5000

DISTRIBUTION: PCN 124 059261 01

CHANGE DEPART- No. 1 } HEADQUARTERS, U.S. MARINE CORPS, MENTS OF THE ARMY, NAVY, AND AIR FORCE WASHINGTON, D.C., 30 June 1989

Organizational ,Intermediate (Field), (Direct and General Support), and Depot Maintenance Repair Parts and Special Tools List

GENERATOR SET, DIESEL ENGINE DRIVEN, TACTICAL, SKID MOUNTED, 3 KW, 3 PHASE 120/208 AND SINGLE PHASE 120/240 VOLTS AC AND 28 VOLTS DC

DOD MODEL	CLASS	MODE	NSN
MEP-016B	UTILITY	60 HZ	6115-01-150-4140
MEP-021B	UTILITY	400 HZ	6115-01-151-8126
MEP-026B	UTILITY	28 VDC	6115-01-150-0367

This publication is required for official use for administrative or operational purposes only. Distribution is limited to U.S. Government agencies. Other re-quests for this document must be referred to:Commandant of the Marine Corps
(HQSP-2) Washington, D.C. 20380-0001.

Destruction Notice:For unclassified, limited documents, destroy by any method that will prevent disclosure of contents or reconstruction of the document.

1.Change 1 to this joint technical manual SL-4-05926B/06509B-24P/2/TM 5-6115-615-24P/NAVFAC P-8-646-24P/TO 35C2-3-386-34 is for Army use only. It is effective upon receipt and provides organizational, direct support, and general support repair parts and special tools list for Generator Set, Engine Driven, Tactical, Skid-Mounted, 3 KW DOD Model MEP 701A, 60 Hz, NSN 6115-01-234-5966. Model 701A is a modified model MEP 016B with the addition of the Acoustic Suppression Kit (ASK), NSN 6115-01-271-1584.The ASK is intended to suppress the high noise level inherent in a diesel-driven generator.

2. Reporting of discrepancies or suggested changes should be submitted on a DA Form 2028 directly to:

Commander
U.S. Army Troop Support Command ATTN : AM-STR-MCTS
4300 Goodfellow Boulevard St. Louis, MO 63120-1798

3. Army Users: Remove and insert the following pages as indicated below:

Remove Pages	Insert Pages
i and ii 164 through 219/220	i and ii 164 through 186

4.Retain this sheet in front of manual for reference purposes.

C1 BY DIRECTION OF THE COMMANDANT OF THE MARINE CORPS

OFFICIAL:

J. G. O'NEILL
Head, Acquisition Support Branch
Acquisition Division
Marine Corps Research, Development and
Acquisition Command

CARL E. VUONO
General, United States Army
Chief of Staff

WILLIAM J. MEEHAN, II
Brigadier General, United States Army
The Adjutant General

B. F. MONTOYA, RADM, CEC
USN

ALFRED G. HANSEN, General, USAF
Air Force Logistics Command

LARRY D. WELSH, General
USAF, Chief of Staff

DISTRIBUTION:
To be distributed in accordance with Da Form 12-25A, Unit, Direct Support and General Support Maintenance requirements for Generator Set, Gas Driven, 120/208V, 120/240V, 28V DC, 3 KW, 60/400HZ, 3 PH (MEP-021A, MEP 026A).

Marine Corps SL-4-05926B/06509B-24P/2
Army TM 5-6115-615-24P
Navy NAVFAC P-8-646-24P
Air Force TO 35C2-3-386-34

DEPARTMENT OF THE NAVY Headquarters, U.S. Marine Corps
Washington, D.C.20380-0001

31 December 1987

1. This Manual is effective upon receipt and contains organizational, intermediate (field) ,(direct and general support) , and depot mainte-nance repair parts and special tools lists for Generator Set, Engine Driven, Tactical, Skid Mounted, 3 KW, DOD Models MEP 016B, 60 HZ, NSN 6115-01-150-4140, MEP 021B, 400 HZ, NSN 6115-01-151-8126, and MEP 026B, 28 VDC, NSN 6,115-01-150-0367.

2.Notice of discrepancies or suggested changes: refer to page 9, and paragraph 7, of this Manual, for reporting errors for applicable Services'form number and forwarding address.

BY DIRECTION OF THE COMMANDANT OF THE MARINE CORPS

OFFICIAL:

J. G. O'NEILL
Head, Acquisition Support Branch
Acquisition Division
Marine Corps Research, Development
and Acquisition Command

CARL E. VUONO
General, United States Army

Brigadier General, United States Army
The Adjutant General

B. F. MONTOYA, RADM, CEC
USN

ALFRED G. HANSEN, General, USAF
Air Force Logistics Command

LARRY D. WELSH, General
USAF, Chief of Staff

DISTRIBUTION :AGB/L77/L82

Copy to:7000161(2)

MARINE CORPS STOCK LIST SL-4-05926B/06509B-24P/2
DEPARTMENT OF THE ARMY TECHNICAL MANUAL TM 5-6115-615-24P
DEPARTMENT OF THE NAVY PIJBLICATION NAVFAC P-8-646-24P
DEPARTMENT OF THE AIR FORCE TECHNICAL ORDER TO 35C2-3-386-34

HEADQUARTERS U.S. MARINE CORPS, DEPARTMENTS OF THE ARMY, NAVY, AND AIR FORCE WASHINGTON, D,C. (DEC 1987)

ORGANIZATIONAL, INTERMEDIATE (FIELD), (DIRECT AND GENERAL SUPPORT), AND DEPOT MAINTENANCE REPAIR PARTS AND SPECIAL TOOLS LIST

GENERATOR SET, DIESEL ENGINE DRIVEN, TACTICAL, SKID MOUNTED, 3 KW, 3 PHASE 120/208 AND SINGLE PHASE 120/240 VOLTS AC AND 28 VOLTS DC

DOD MODEL	CLASS	MODE	NSN
MEP-016B	UTILITY	60 HZ	6115-01-150-4140
MEP-021B	UTILITY	400 HZ	6115-01-151-8126
MEP-026B	UTILITY	28 VDC	6115-01-150-0367

Current as of December 1987

TABLE OF CONTENTS

TABLE OF CONTENTS - Continued

LIST OF EFFECTIVE PAGES

INSERT LATEST CHANGED PAGES. DESTROY SUPERSEDED PAGES.

NOTE: The portion of the text affected by the changes is indicated by a vertical line in the outer margins of the page. Changes to Illustrations are indicated by miniature pointing hands. Changes to wiring diagrams are indicated by shaded areas.

Date of issue for original and changed pages are:
Original...0... December 1987

TOTAL NUMBER OF PAGES IN THIS PUBLICATION IS 176 CONSISTING OF THE FOLLOWING :

Page No.	* Change No.
Cover	0
Blank	0
A.	0
Blank i	0
thru ii . . . . 1 thru	0
18 . . . . Blank . . .	0
. . . . 20 thru 36 . . .	0
. . . . Blank	0
. . . . 38 thru 44	0
. . . . Blank	0
. . . . 46 thru 50	0
Blank 52	0
thru 56 . . . . Blank .	0
. . . . 58 thru 60 .	0
. . . . Blank	0
. . . .	0
	0

Page No.	* Change No.
62 thru 72	0
Blank	0
74 thru 114	0
Blank	0
116 thru 124	0
Blank	0
126 thru 138	0
Blank	0
140 thru 148	0
Blank	0
150 thru 152	0
Blank	0
154 thru 158	0
Blank	0
160 thru 162	0
Blank	0

* Zero in this column indicates an original page.

ORGANIZATIONAL, INTERMEDIATE (FIELD) (DIRECT SUPPORT AND GENERAL SUPPORT) AND DEPOTMAINTENCE REPAIR PARTS AND SPECIAL TOOLS LIST

Section I. INTRODUCTION

1.SCOPE. This joint Army, Navy, Air Force and Marine Corps manual lists repair parts and special tools required for the performance of organizational, intermediate (field) (direct and general support) and depot maintenance of the generator set.The following paragraphs are keyed to applicable users.All users should read paragraph 4, Special Information, prior to using this manual.

2. GENERAL.This Repair Parts and Special Tools List is divided into the following sections:

3. (ALL) Repair Parts - Section II. A list of repair parts authorized for the performance of maintenance at the organizational, intermediate (field) (direct support and general support) and depot level in figure and item number sequence.

b.(ALL) Special Tools, Test and Support Equipment - Section III. A list of special tools, test and support equipment authorized for the performance of maintenance at the organizational, intermediate (field) (direct support and general support) and depot level.

c. National Stock Number and Reference Number Index - Section IV. A list of National stock numbers in numerical sequence, followed by a list of reference numbers appearing in all the listings, in alphanumeric sequence, cross referenced to the illustration figure number and item number.

d.Reference Designator Index -Section V.The reference designator column includes all assigned reference designators arranged first in alphabetical order, second in numeric order. Opposite each symbol is listed the figure and item number of the part in Section II and the reference number.

3.EXPLANATION OF COLUMNS. The following provides an explanation of columns in the tabular lists in Sections II and III.

a. (ALL) Illustrations. (column I)This column is divided as follows:

(1) Figure Number. Indicates the figure number of the illustration on which the item is shown.

(2) Item Number. Indicates the number used to identify the item on the illustration.

b. (ALL) Source, Maintenance, and Recoverability Codes (SMR).(Column 2)

(1) Uniform source codes applicable to all Military Services.

GENERAL: Source codes are assigned to support items to indicate the manner of acquiring support items for maintenance, repair, or overhaul of end items. Source codes are entered in the first and second positions of the Uniform SMR Code format as follows:

Code	Definition
PA	Item procured and stocked for anticipated or known usage.
PB	Item procured and stocked for insurance purposes because essentiality dictates that a minimum quantity be available in the supply systems.

Pc Item procured and stocked and which otherwise would be coded PA except that it is deteriorative in nature.

PD Support item, excluding support equipment, procured for initial issue or outfitting and stocked only for subsequent or additional initial issues or outfitting.Not subject to automatic replenishment.

PE Support equipment procured and stocked for initial issue or outfittings to specified maintenance repair activities.

PF Support equipment which will not be stocked but which will be centrally procured on demand.

PG Item procured and stocked to provide for sustained support for the life of the equipment. It is applied to an item peculiar to the equipment which because of probable discontinuance or shutdown of production facilities would prove uneconomical to reproduce at a later time.

KD An item of depot overhaul/repair kit and not purchased separately. Depot kit defined as a kit that provides items required at the time of overhaul or repair.

KF An item of maintenance kit and not purchased separately. Maintenance kit defined as a kit that provides an item that can be replaced at organizational or intermediate levels of maintenance.

KB Item included in both a depot overhaul repair kit and a maintenance kit.

MO Item to be manufactured or fabricated at organizational level.

MF Item to be manufactured or fabricated at the following intermediate maintenance levels:

USAF -Intermediate (*)
USA- Direct Support (*)
USMC - Third Echelon
USN -Ashore

MG Item to be manufactured or fabricated at both afloat and ashore intermediate maintenance levels - Navy use only.

MH Item to be manufactured or fabricated at the general support maintenance level.

MD Item to be manufactured or fabricated at depot maintenance level.

AO Item to be assembled at organizational level.

AF Item to be assembled at the following intermediate

USAF - Intermediate (*)
USA - Direct Support (*)
USMC- Third Echelon
USN -Afloat

AH Item to be assembled at the following intermediate maintenance levels:

USAF - Intermediate (*)
USA- General Support (*) USMC - Fourth Echelon
USN -Ashore

AG Item to be assembled at both afloat and ashore intermediate maintenance levels - Navy use only.

AD Item to be assembled at depot maintenance level.

XA Item is not procured or stocked because the requirements for the item will result in the replacement of the next higher assembly.

XB Item is not procured or stocked. If not available through salvage, requisition.

XC Installation drawings, diagram, instruction sheet, field service drawing that is identified by manufacturer's part number.

XD A support item that is not stocked. When required item will be procured through normal supply channels.

(*) NOTE: For USAF and the,USA Safeguard Program.Only Code "F" will be used to denote intermediate maintenance. On joint programs, use of either code F or H by the joining service will denote intermediate maintenance to USAF and USA Safeguard Program.

(2) Uniform maintenance codes applicable to all Military Services:

GENERAL :Maintenance codes are assigned to indicate the levels of maintenance authorized to USE and REPAIR support items.The maintenance codes are in the third and fourth position of the Uniform SMR Code Format.

(a) USE (THIRD POSITION): The maintenance code entered in the third position indicates the lowest maintenance level authorized to remove replace, and use the support item.The maintenance code entered in the third position indicates one of the following levels of maintenance:

Code **Application/Explanation**

oSupport item is removed, replaced, used at the organizational level of maintenance.

F Support item is removed, replaced, used at the following intermediate levels:

USAF -Intermediate (*)
USA - Direct Support (*)
USN --Afloat
USMC- Third Echelon

G Support item is removed, replaced, used at both afloat and ashore intermediate Levels- Navy use only.

H Support item is removed, replaced, used at the following intermediate levels:

USAF -Intermediate (*)
USA - General Support (*)
USN -Ashore (only)
USMC -Fourth Echelon

D Support items that are removed, replaced, used at depot only:

USAF -Depot.Mobile Depot and Specialized RepairActivity USA -Depot. Mobile Depotand Specialized RepairActivity USN -Aviation Rework.Avionics and Ordinance Facilities and Shipyards
USMC -Depot

(b) REPAIR (FOURTH POSITION):

The maintenance code entered in the fourth position indicates whether the item is to be repaired and identifies the lowest maintenance level with the capability to perform complete repair (i.e., all authorized maintenance functions).

Code Application/Explanation

o The lowest maintenance level capable of complete repair of the support item is the organizational level.

F The lowest maintenance level capable of complete repair of the support item is the following intermediate levels:

USAF -Intermediate (*)
USA- Direct Support (*)
USN- Ashore (only)
USMC -Third Echelon

H The lowest maintenance level capable of complete repair of the support item is the following intermediate levels:

USAF -Intermediate (*)
USA- General Support (*)
USN - Ashore (only)
USMC -Fourth Echelon

(*) NOTE: For USAF programs and the USA Safeguard Program Code F will be used to denote intermediate maintenance.On joint programs, use of either Code F or H by the joining Service will denote intermediate maintenance to USAF and the USA Safeguard Program.

GBoth afloat and ashore intermediate levels are capable of complete repair of support item -Navy use only.The lowest maintenance level capable of complete repair of the support item is the depot level.

USAF - Depot. Mobile Depot, and Specialized Repair Activity
USA - Depot. Mobile Depot and Specialized Repair Activity
USN -Aviation Rework. Avionics and Ordanance Facilities Shipyards
USMC - Depot.

LRepair restricted to designated Specialized Repair Activity.

zNonreparable. No repair is authorized.

BNo repair is authorized. The item may be reconditioned by adjusting, lubricating, etc., at the user level. No parts or special tools are procured for the maintenance of this item.

(3) Uniform Recoverability Codes applicable to all Military Services.

GENERAL: Recoverability codes are assigned to support items to indicate the disposition action on unserviceable items. The recoverability code is entered in the fifth position of the uniform SMR Code Format as follows:

Code	Definition
z	Nonreparable item.When unserviceable, condemn and dispose at the level indicated in column 3.
o	Reparable item. When uneconomically reparable, condemn and dispose at organizational level.
F	Reparable item.When uneconomically reparable, condemn and dispose at the following intermediate levels: USAF -Intermediate (*) USA - Direct Support (*) USN - Afloat USMC- Third Echelon
H	Reparable item. When uneconomically reparable, condemn and dispose at the following levels: USAF -Intermediate (*) USA- General Support (*) USN -Ashore USMC -Fourth Echelon

(*) NOTE: For USAF programs and the USA Safeguard Program, Code F will be used to denote intermediate maintenance. On joint programs, use of either Code F or H by the joining Service will denote intermediate level of USAF and the USA Safeguard Program.

DReparable item. When beyond lower level repair capability, return to depot. Condemnation and disposal not authorized below depot level.

L Reparable item.Repair, condemnation and disposal not authorized below depot/Specialized Repair Activity level.

A Item requires special handling or condemnation procedure because of specific reason (i.e., precious metal content, high-dollar value, critical material or hazardous material).Refer to appropriate manuals/directives for specific instructions.

c. (All) National Stock Number

(NSN) .(Column 3). Indicates the NSN assigned to the item and will be used for requisitioning purposes.

d. (ALU Description. (Column 4).Indicates the Federal item name and any additional descriptions of the item required. The abbreviation"w/e" when used as part of the nomenclature, indicates that the NSN includes all armement, equipment, accessories and repair parts issued with the item. A part number or other reference number is followed by the applicable five digit FSCM in parentheses. If two reference numbers and Federal Supply Codes for manufacturer are listed, the first listing refers to the DOD Drawing Number, the second listing refers to the actual part manufacturer.Items that are included in kits and sets are listed below the name of the kit or set with the quantity of each item in the kit or set indicated in the quantity incorporated in unit column.

e. (ALL) Unit of Measure (U/M). (Column 5).Indicates the standard of the basic quantity of the listed item as used in performing the actual maintenance function. This measure is expressed by a two-character alphabetical abbreviation (e.g., ea. in. pr. etc.). When the unit of measure differs from the unit of issue, the lowest unit of issue that will satisfy the required units of measure will be requisitioned.

f.(ALL) Quantity Incorporated unit in (Column 6).Indicates the qumtity of the item used in the assembly group. A "V" appearing in this column in lieu of a quantity indicates that a definite quantity cannot be indicated (e.g., shims, spacers, etc.).

g. (MC) Quantity Per Equipment. (Column 7).Indicates the consolidated quantity of a maintenance part used on an end item upon the parts first appearance in this publication.

4.SPECIAL INFORMATION.

a. (ALL) Usable On Code.

Identification of the usable on codes of this publication for various DOD

code	Used On
A	MEP-016B
B	MBP-026B
C	NBP-021B

b.(Army) Higher Level Code. Organizational maintenance personnel will extract the items which they require from Section II, 3rd or 4th position of column 2 of the direct and general support RPSTL. Parts which are manufactured or assembled at a higher level than that authorized to install the part are indicated by the use of higher level code in the source column.

c. Stockage Information.

(1) Air Force stockage information is contained in Initial Supply Support Lists issued separately from this publication by Sacramento Air Logistics Center in accordance with AFM 67-1. part 1. chapter 12.

(2) Navy stockage objectives are contained in Consolidated Shipboard Allowance List as furnished separately from this publication by the Ship Parts Control Center.

(3) Army stockage is demand based in accordance with AR 710-2. Repair parts listed in this publication represent those authorized for use at indicated maintenance levels and will be requisitioned on an as required basis until stockage is justified in accordance with AR 710-2.

5. HOW TO LOCATE REPAIR PARTS.

a. (ALL) When NSN or reference number is unknown:

(1) Using the table of contents, determine the functional group: i.e., batteries and related parts, exhaust and breather pipes, within which the repair part belongs. This is necessary since illustrations are prepared for functional groups, and listings are divided into the same groups.

(2) Find the illustration covering the functional group to which the repair part belongs.

(3) Identify the repair part on the illustration and note the illustration figure and item number of the repair part.

(4) Using the Repair Parts Listing, find the figure and item number noted on the illustration.

b. (ALL) When NSN or reference number is known:

(1) Using the NSN and Reference Numbers Index, find the pertinent NSN or reference number. This index is in ascending NSN sequence followed by a list of reference numbers in alphanumeric sequence, cross-referenced to the illustration figure number and item number.

(2) After finding the figure and item number, locate the figure and item number in the repair parts list.

6. (F) USE OF THE REFERENCE DESIGNATOR INDEX SECTION. This Section (Section V) is used when the reference designator is known or identified by other technical manuals supporting this equipment. The reference number is given in this Section. If description or location is desired, note the figure and item number. Turn to Section II to the noted figure and item number. The location of the part and description is given in this listing.

7. ABBREVIATIONS .

Abbreviation	**Explanation**
(All abbreviations are covered under MIL-STD-12)	

8. COMMERCIAL AND GOVERNMENT **ENTITY (CAGE) CODE.**

Code	Manufacturer
19207	U.S. Army Tank Automotive Command Warren, MI 48090
30554	Project Manager - Mobile Electric Power 7500 Backlick Road Springfield, VA 22150-3107
44940	ONAN Corp.U.S. Power Product Div. 1400 73rd Ave. N.E. Minneapolis, MN 55428
45722	USM Corp. Subsidiary of Emhart Industries, Inc. Parker-Kalon Fastener Div. Campbellsville, KY 42718
70411	Anderson Brass Co. 100 S. Campbell Ave. Detroit, MI 48209-3102
72850	Facet EnterPrises Motor Component Div. 18Th Street and Oakwood Ave. Elmira, NY 14903

76700 Nelson Div. Nelson Industries Inc.
Hwy. 51 West P.O. Box 428
Stoughton, WI 53589

81348 Federal Specifications

81349 Military Specifications

88044 Aeronautical Standards Group
Dept. of Navy and Air *Force*

91349 Missouri Brush and Crayon Co.
St. Louis, MO

96906 Military Standards

97403 U.S. Army Belvoir Research and Development Center
Fort Belvoir, VA 22060

9. RECOMMENDATION OF MAINTENANCE PUBLICATION IMPROVEMENTS.
Report of errors, omissions and recommendations for improving this publication by the individual user is encouraged.Reports should be submitted as follows:

a. Air Force. AFTO Form 22 in accordance with T.O. 00-5-1, directly to:Commander Sacramento Air Logistics Center, ATTN MMEDT, McClellan AFB, CA 95652

b. Army. DA Form 2028, directly to: Commander, .US Army Troop Support Command, ATTN: AMSTR-MCTS, 4300 Goodfellow Boulevard, St. Louis, MO 63120

C. Marine Corps. NAVMC 10772 directly to: Commanding General, Marine Corps Logistics Base (Code 850), Albany, Georgia 31704-5000.

d. Navy. Letter, directly to: Commanding Officer, US Navy Ship Parts Control Center, ATTN: Code 783, Mechanicsburg, PA 17055

Section II. REPAIR PARTS LIST

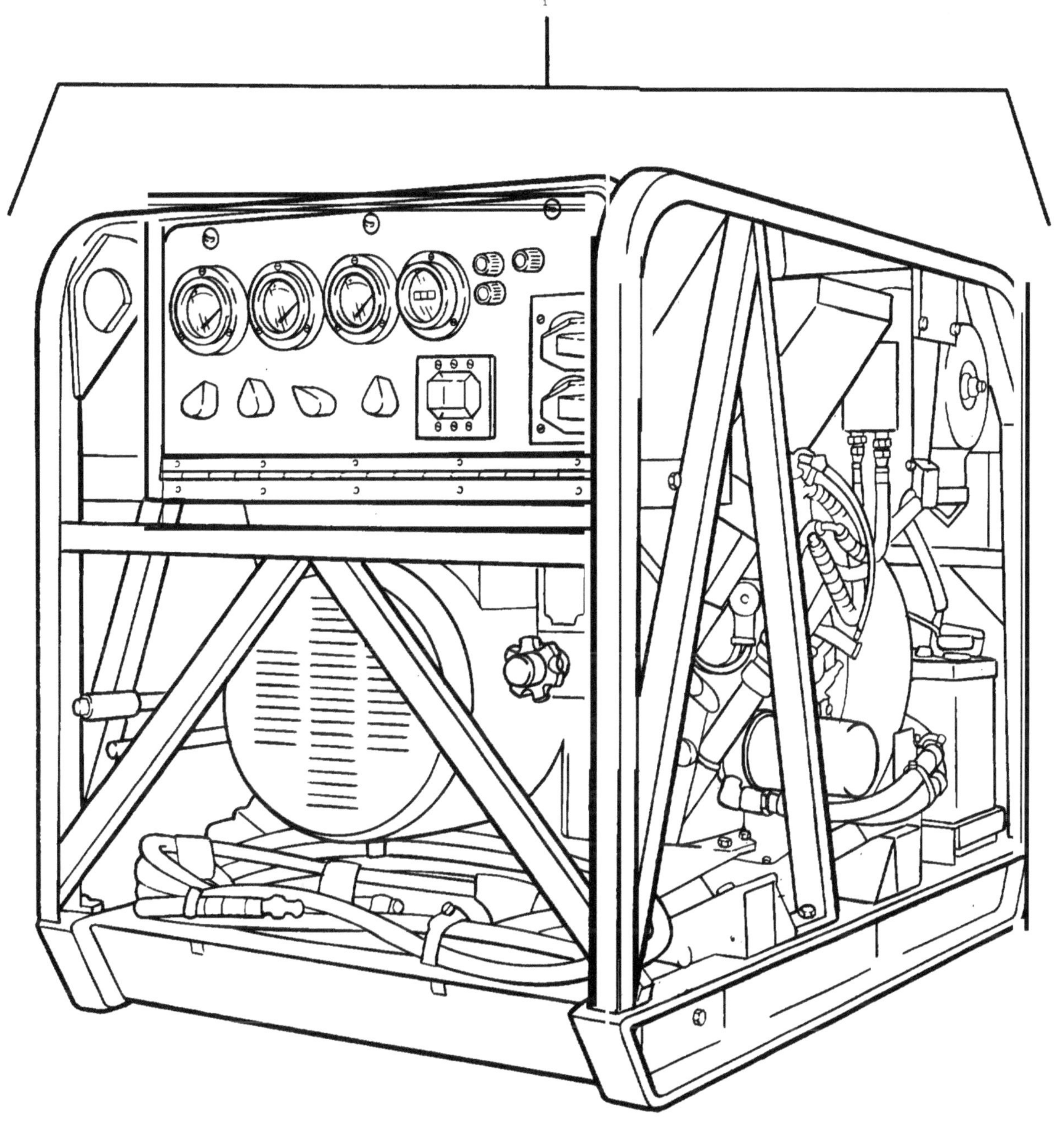

Figure 1. Generator Set, 60 Hz

(1) ILLUSTRATION (a) FIG NO.	(b)	(2) SMR CODE a ARMY	b AIR FORCE	d NAVY	e USMC	(3) NATIONAL STOCK NUMBER	(4) DESCRIPTION REF NUMBER MFR CODE	USABLE ON CODE	(5) U/M	(6) QTY INC IN	(7) USMC QTY PER EQUIP
1	1	PDOHH	PDOHH	PDOHH	PDOHH	6115401-1504140	GENERATOR SET, DIESEL MEP-016B (30554)	A	EA	1 1	

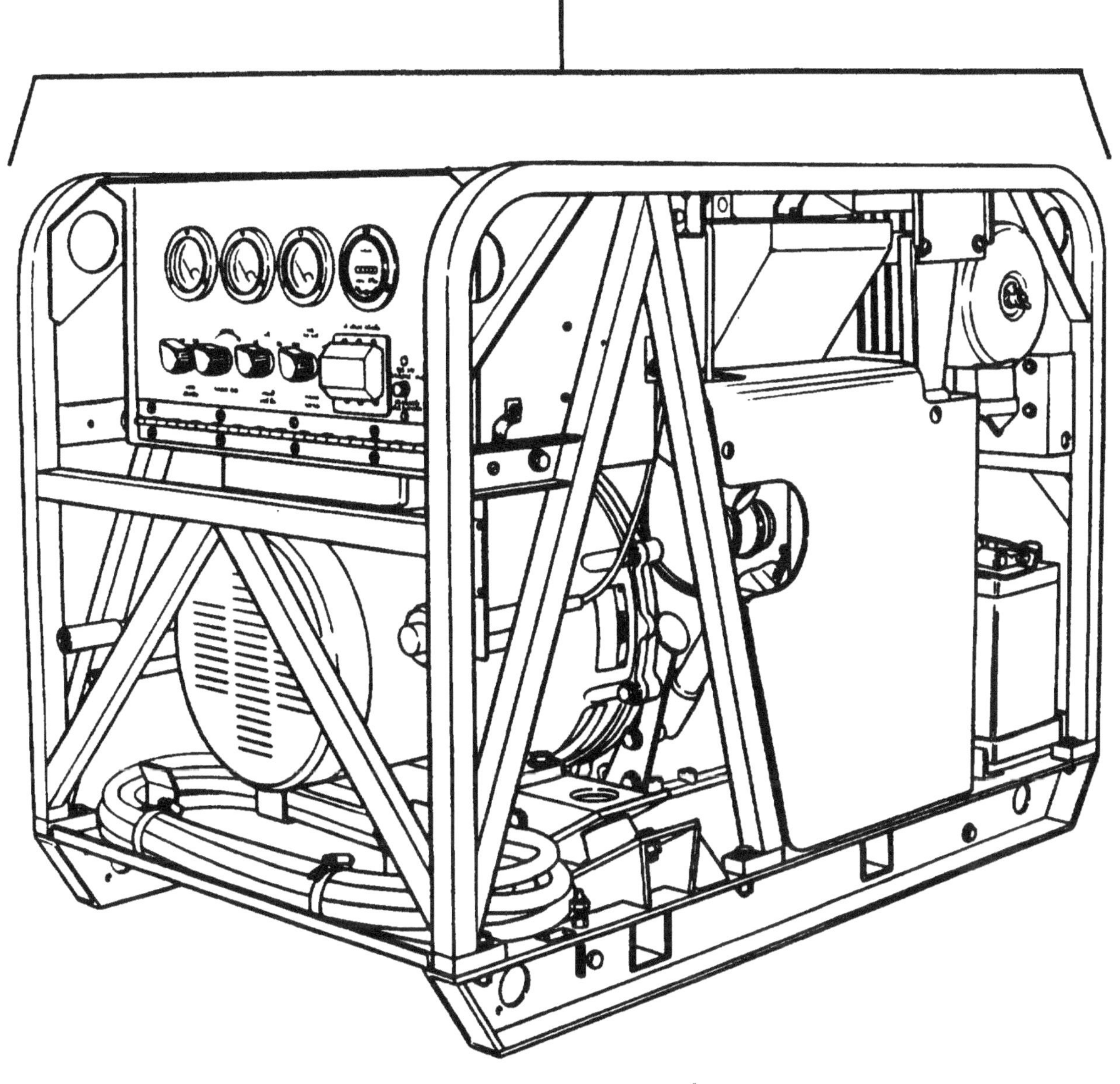

Figure 2. Generator Set, 400 Hz

(1) ILLUSTRATION		(2) SMR CODE				(3)	(4)	(5)	(6)	(7)
(a) FIG NO.	(b)	a ARMY	b AIR FORCE	d NAVY	e USMC	NATIONAL STOCK NUMBER	DESCRIPTION USABLE ON CODE REF NUMBER MFR CODE	U/M	QTY INC IN	USMC QTY PER EQUIP
2	1	PDOHH	PDOHH	PDOHH	PDOHH	6115-01-151-8126	GENERATOR SET, DIESEL .... B MEP-021B (30554)	EA 1	1	

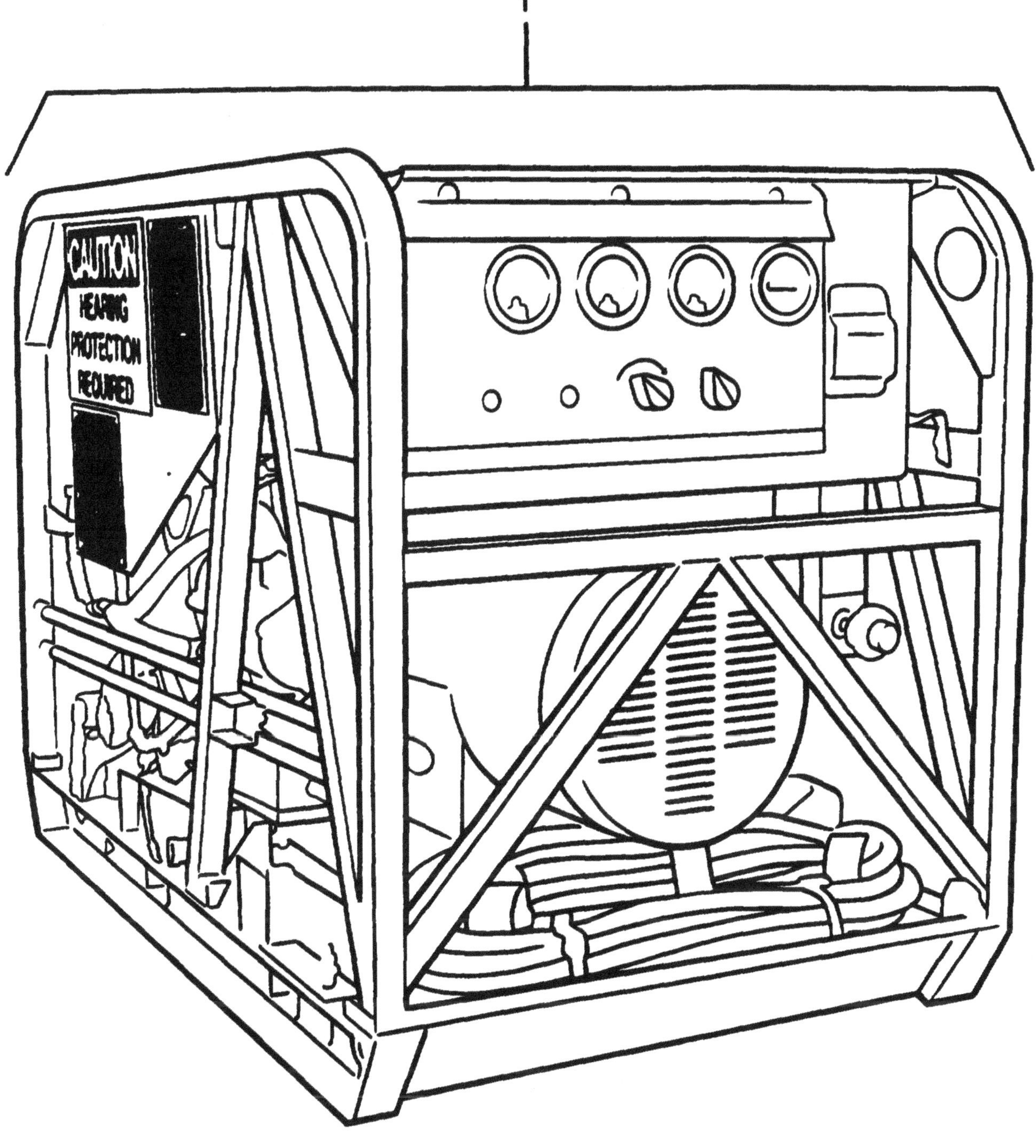

Figure 3. Generator Set, 28 VDC

(1) ILLUS-TRA-TION		(2) SMR CODE				(3)	(4)	(5)	(6)	(7)
(a) FIG NO.	(b)	a ARMY	b AIR FORCE	d NAVY	e USMC	NATIONAL STOCK NUMBER	DESCRIPTION USABLE ON CODE REF NUMBER MFR CODE	U/M	QTY INC IN	USMC QTY PER EQUIP
3	1	PDOHH	PDOHH	PDOHH	PDOHH	6115-01-150-0367	GENERATOR SET, DIESEL .... C MEP-026B (30554)	EA 1	1	

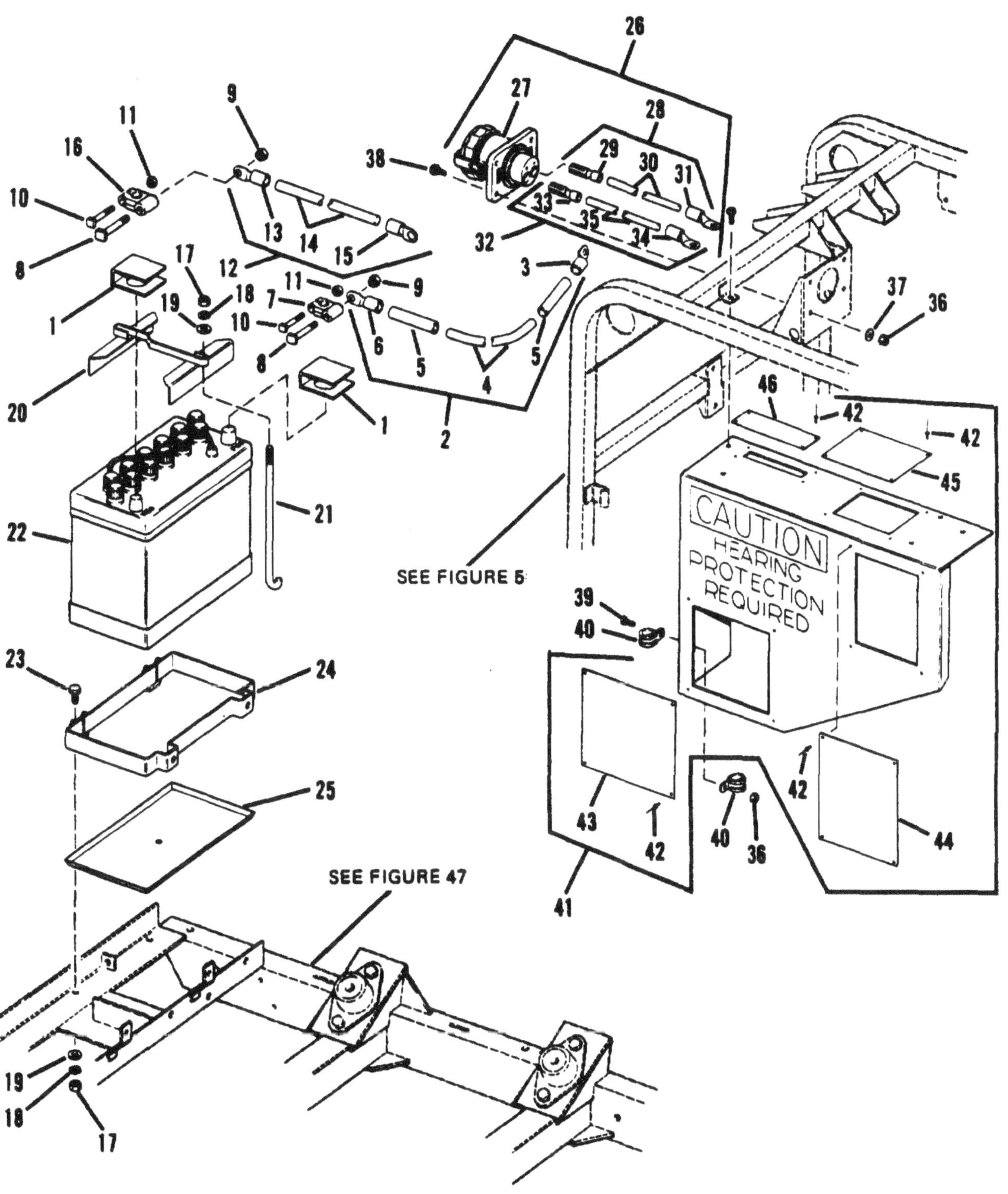

Figure 4. Battery, Battery Cables and Receptacle, Group 01 (DC Electrical and Control System)

(1) ILLUSTRATION (a) FIG NO.	(1) (b)	(2) SMR CODE a ARMY	b AIR FORCE	d NAVY	e USMC	(3) NATIONAL STOCK NUMBER	(4) DESCRIPTION REF NUMBER MFR CODE	USABLE ON CODE	(5) U/M	(6) QTY INC IN	(7) USMC QTY PER EQUIP
							GROUP 01 DC ELECTRICAL AND CONTROL SYSTEM				
4	1	PAOZZ	PAOZZ	PAOZZ	PAOZZ	5940-00-738-6272	COVER BATTERY TERMINAL.............. 10942521 (19207)	ABC	EA 2	2	
4	2	AFOOO	AFOOO	AFOOO	AFOOO	6150-01-286-9552	LEAD, ELECTRICAL 416-0857 (4484) 84-13048 (30554)	ABC	EA 1	1	
4	3	PAOZZ	PAOZZ	PAOZZ	PAOZZ	5940-01-054-6954	TERMINAL, LUG 326 896 (00779 72-15319 (30554)	ABC	EA 1	1	
4	4	PAOZZ	PAOZZ	PAOZZ	PAOZZ		WIRE, ELECTRICAL MIL-C-5756 (81349)	ABC			
4	5	PAOZZ	PAOZZ	PAOZZ	PAOZZ	6145-01-054-6954	SLEEVING, INSULATION..................... M23053/5-109-2 (81349	ABC	IN 90	38	
4	6	PAOZZ	PAOZZ	PAOZZ	PAOZZ	5970-00-914-3117	TERMINA, LUG MS25036-127 (96906)	ABC	IN 6	6	
4	7	PAOZZ	PAOZZ	PAOZZ	PAOZZ	5940-00-113-8191	TERMINAL, LUG MS75004-1 (96906)	ABC/	EA	1	4
4	8	PAOZZ	PAOZZ	PAOZZ	PAOZZ	5940-00-549-6581	BOLT, MACHINE.................................. MS3535667 (96906)	ABC	EA 1	1	
4	9	PAOZZ	PAOZZ	PAOZZ	PAOZZ		NUT, PLAIN, HEXAGON MS51967-8 (96906)	ABC	EA 2	2	
4	10	PAOZZ	PAOZZ	PAOZZ	PAOZZ	5306-01-280-6685	BOLT, MACHINE.................................. MS35356-35 (96906)	ABC	EA 2	2	
4	11	PAOZZ	PAOZZ	PAOZZ	PAOZZ	5310-00-732-0558	NUT, PLAIN, HEXAGON MS51967-5 (96906)	ABC			
4	12	PAOOO	PAOOO	PAOOO	PAOOO	5306-01-276-0837	LEAD ELECTRICAL 416-0856 (44940) 84-13049 (30554)	ABC	EA	2	2
4	13	PAOZZ	PAOZZ	PAOZZ	PAOZZ	5310-00-880-7744	TERMINAL, LUG MS25036-127 (96906)	ABC	EA 2	2	
4	14	PAOZZ	PAOZZ	PAOZZ	PAOZZ	6150-01-280-0453	WIRE, ELETRICIAL.............................. MIL-C-5756 (81349)	ABC	EA 1	1	
4	15	PAOZZ	PAOZZ	PAOZZ	PAOZZ		TERMINAL LUG MS25036-128 (96906)	ABC			
4	16	PAOZZ	PAOZZ	PAOZZ	PAOZZ	5940-00-113-8191	TERMINAL LUG MS75004-2 (96906)	ABC	EA 1	1	
4	17	PAOZZ	PAOZZ	PAOZZ	PAOZZ	6145-01-047-0530	NUT, PLAIN, HEXAGON MS51967-6 (96906)	ABC	IN 27	27	
									EA	1	1
						5940-00-113-					
									EA	1	1

(1) ILLUSTRATION (a) FIG NO.	(1) (b)	(2) SMR CODE a ARMY	b AIR FORCE	d NAVY	e USMC	(3) NATIONAL STOCK NUMBER	(4) DESCRIPTION / REF NUMBER MFR CODE	USABLE ON CODE	(5) U/M	(6) QTY INC IN	(7) USMC QTY PER EQUIP
							GROUP 01 DC ELECTRICAL AND CONTROL SYSTEM				
4	18	PAOZZ	PAOZZ	PAOZZ	PAOZZ	5310-00-407-9566	WASHER, LOCK MS35338-45 (96906) 850-1045 (44940)	ABC	EA	6	
4	19	PAOZZ	PAOZZ	PAOZZ	PAOZZ	5310-00-081-4219	WASHER, FLAT MS27183-12 (96906)	ABC	EA	6	67
4	20	XBOZZ	XBOZZ	XBOZZ	XBOZZ		BATTERY FRAME, TOP 416-0859 (49940) 84-13078 (30554)	ABC	EA	1	1
4	21	XBOZZ	XBOZZ	XBOZZ	XBOZZ		BOLT, HOOK..... 406-0584 (49940) 84-13078 (30554)	ABC	EA	2	2
4	22	PAOZZ	PAOZZ	PAOZZ	PAOZZ		BATTERY, STORAGE..... MS750047-1 (96906)	ABC	EA	1	1
4	23	PAOZZ	PAOZZ	PAOZZ	PAOZZ	6140-00-059-3528	BOLT, MACHINE..... MS90725-31 (96906)	ABC	EA	4	12
4	24	XBOZZ	XBOZZ	XBOZZ	XBOZZ	5306-00-225-8496	BATTERY FRAME, BOTTOM..... 416-0860 (49940) 84-13076 (30554)	ABC	EA	1	1
4	25	XBOZZ	XBOZZ	XBOZZ	XBOZZ		TRAY, BATTERY..... 416-0858 (49940) 84-13061 (30554)	ABC	EA	1	1
4	26	AFOOO	AFOOO	AFOOO	AFOOO		RECEPTACLE, SLAVE 323-1326 (44940) 84-13064 (30554)	ABC	EA	1	1
4	27	PAOZZ	PAOZZ	PAOZZ	PAOZZ		CONNECTOR, RECEPTACLE MS75058-1 (96906)	ABC	EA	1	1
4	28	AFOOO	AFOOO	AFOOO	AFOOO	5935-00-295-6403	LEAD ASSY, NEGATIVE..... 226-3079 (44940) 84-13063 (30554)	ABC	EA	1	1
4	29	PAOZZ	PAOZZ	PAOZZ	PAOZZ		CONNECTOR, PLUG 1162338 (19207)	ABC	EA	1	1
4	30	PAOZZ	PAOZZ	PAOZZ	PAOZZ	5935-00-567-0128	WIRE, ELECTRICAL MIL-C-5756 (81349)	ABC	IN	14	14
4	31	PAOZZ	PAOZZ	PAOZZ	PAOZZ	6145-01-047-0530	TERMINAL, LUB..... MS25036-127 (96906)	ABC	EA	1	1

(1) ILLUSTRATION (a) FIG NO.	(b)	(2) SMR CODE a ARMY	b AIR FORCE	d NAVY	e USMC	(3) NATIONAL STOCK NUMBER	(4) DESCRIPTION REF NUMBER MFR CODE / USABLE ON CODE	(5) U/M	(6) QTY INC IN	(7) USMC QTY PER EQUIP
							GROUP 01 DC ELECTRICAL AND CONTROL SYSTEM			
4	32	AFOOO	AFOOO	AFOOO	AFOOO		LEAD ASSY, POSITIVE .... ABC 226-307 (44940) 84-130 (30554)	EA 1	1	
4	33	PAOZZ	PAOZZ	PAOZZ	PAOZZ	593540-567-0128	CONNECTOR, PLUG .... ABC 11682338 (19207)	EA 1	1	
4	34	PAOZZ	PAOZZ	PAOZZ	PAOZZ	594040-1134191	TERMINAL, LUG .... ABC MS2503627 (96906)			
4	35	PAOZZ	PAOZZ	PAOZZ	PAOZZ	6145-01-04730	WIRE, ELECTRICAL .... ABC MIL-C5756 (81349)	EA 1	1	
4	36	PAOZZ	PAOZZ	PAOZZ	PAOZZ	531040596173	NUT, PLAIN, ASSEMBLED .... ABC 501-250800-00 (78189) 69-561-5 (30554)	IN 11	11	
4	37	PAOZZ	PAOZZ	PAOZZ	PAOZZ	53104.8094058	WASH FLAT .... ABC MS27183-10 (96906)	EA	7 34	
4	38	PAOZZ	PAOZZ	PAOZZ	PAOZZ	5305-01-147-8224	SCREW, ASSEMBLED WASHER .... ABC 2AN44 (45152) 294142 (06853)	EA	4 24	
4	39	PAOZZ	PAOZZ	PAOZZ	PAOZZ	5305-01-78-5064	SCREW, ASSEMBLED .... ABC P-15121-78 (45722) 69-662-78 (30554)	EA 4	4	
4	40	PAOZZ	PAOZZ	PAOZZ	PAOZ	5340.404)-1254	CLAMP, LOOP .... ABC MS21333-104 (96916)	EA	7 32	
4	41	XBOZZ	XBOZZ	XBOZZ	XBOZZ		PANEL, ENGINE GUARD .... ABC 403-2374 (44940) 84-13084 (30554)	EA 4	4	
4	42	PAFZZ	PAOZZ	PAOZZ	PAOZZ	5320-01-049-8263	RIVET, BLIND....................................... ABC RV630-4-2 (53551) 818-0076 (44940) 72-5018 (30554)	EA 1	1	
4	43	MDFZZ	MDOZZ	MDOZZ	MDOZZ		PLATE, INFO, LFT, TOW .... ABC 099-2317 (44940) 4-13037 (30554)	EA	16	22
4	44	MDFZZ	MDOZZ	MDOZZ	MDOZZ		PLATE,INFO,FUEL............................... ABC 099-2313 (44940) 84-13038 (30554)	EA 1	1	
4	45	MDFZZ	MDOZZ	MDOZZ	MDOZZ		PLATE, IDENT GEN SET .... A 099-2319-1 (44940) 4-13042 (30544)	EA 1	1	
								EA 1	1	

(1) ILLUSTRATION FIG NO	(2) ITEM NO	(3) SMR CODE ARMY	AIR FORCE	NAVY	USMC	(4) NATIONAL STOCK NUMBER	(5)(6)(7) DESCRIPTIION REF NUMBER MFR CODE	USABLE ON CODE	U/M	QTY INC IN UNITS	USMC QTY PER EQUIP
							GROUP 01 DC ELECTRICAL AND CONTROL SYSTEM				
4	45	MDFZZ	MDOZZ	MDOZZ	MDOZZ		PLATE,IDENT GEN SET 099-2319-2 (44940) 84-13042 (30554)	B	EA	1	1
4	45	MDFZZ	MDOZZ	MDOZZ	MDOZZ		PLATE,IDENT GEN SET 099-2139-3 (44940) 84-13042 (30554)	C	EA	1	1
4	46	MDFZZ	MDOZZ	MDOZZ	MDOZZ		PLATE,INSTR.BTRY 099-2316 (44940) 84-13014 (30554)	ABC	EA	1	1

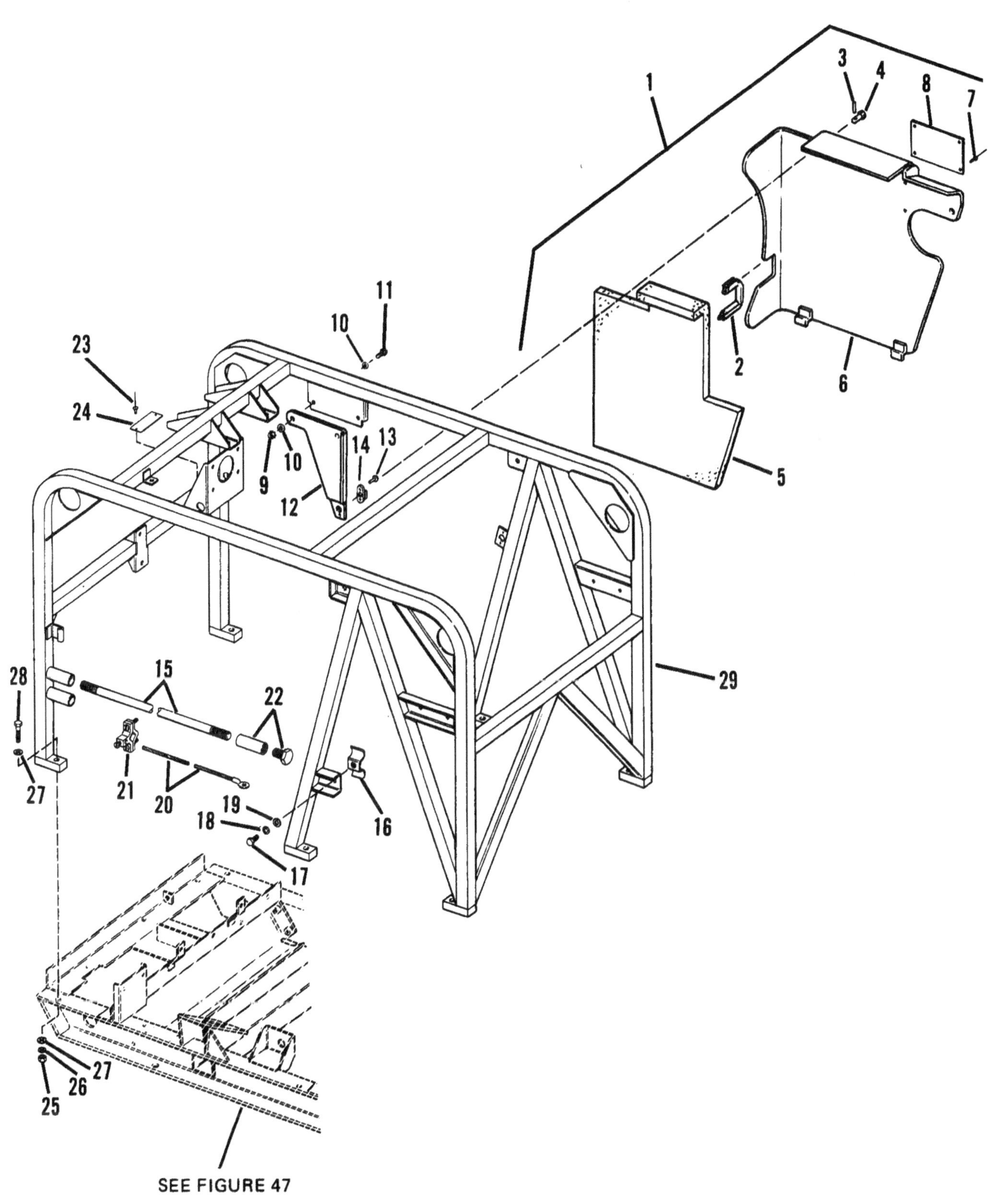

Figure 5. Frame, Ground Assembly and Guard, **Group 02 (Frame)**

(1) ILLUSTRATION (a) FIG NO.	(1) ILLUSTRATION (b)	(2) SMR CODE a ARMY	(2) SMR CODE b AIR FORCE	(2) SMR CODE d NAVY	(2) SMR CODE e USMC	(3) NATIONAL STOCK NUMBER	(4) DESCRIPTION REF NUMBER MFR CODE / USABLE ON CODE	(5) U/M	(6) QTY INC IN	(7) USMC QTY PER EQUIP
							GROUP 02 FRAME			
5	1	XBOZZ	XBOZZ	XBOZZ	XBOZZ		GUARD, ENGINE ASSY..........................AB 134461601 (44940) 84-13225-01 (30554)	EA 1	1	
5	1	XBOZZ	XBOZZ	XBOZZ	XBOZZ		GUARD.ENGINE ASSY.............................. C 134 46 02 (44940) 84-13225 (30554)	EA 1	1	
5	2	PAOZZ	PAOZZ	PAOZZ	PAOZZ	9390-01-287-8896	NONMETALLIC CHANNEL............ ABC 100-1/16-B-3-ALUM (57137) 84-13317 (30554)	FT 1	1	
5	3	PAOZZ	PAOZZ	PAOZZ	PAOZZ	531500449-2945	PIN, GROOVED............................ ABC 99836 (60119) 1005169 (18876) 69-695 (30554)	EA 8	2	
5	4	PAOZZ	PAOZZ	PAOZZ	PAOZZ	5325-00-432-9899	STUD, TURNLOCK........................ ABC 98292-2-220 (60119) MIL-F-5591 (81349)			
5	5	PAOZZ	PAOZZ	PAOZZ	PAOZZ	5330-01-275-3357	GASKET .... ABC 45-3905 (44940) 8413224 (30554)	EA	2 11	
5	6	XBOZZ	XBOZZ	XBOZZ	XBOZZ		PANEL, ENGINE GUARD................AB 134-46 (44940) 5-13220 (30554)	EA 1	1	
5	7	PAOZZ	PAOZZ	PAOZZ	PAOZZ		RIVET, BLIND..........................AB M24243/1-D403 (81349)	EA 1	1	
5	8	MDOZZ	MDOZZ	MDOZZ	MDOZZ	5320-00-395-6523	PLATE, INFORMATION...........AB 13208E5841 (97403)			
5	6	XBOZZ	XBOZZ	XBOZZ	XBOZZ		PANEL, ENGINE GUARD.................. C 134-4687 (44940) 84-13339 (30554)	EA	4 30	
								EA 1	1	
5	9	PAOZZ	PAOZZ	PAOZZ	PAOZZ		NUT, PLAIN, ASSEMBLED .... ABC 501-250800-00 (78189) 69-561-5 (30554)			
								EA 1	1	
5	10	PAOZZ	PAOZ	PAOZZ	PAOZZ	5310-00-696-5173	WASHER.FLAT .... ABC MS27183-10 (96906)			
5	11	PAOZZ	PAOZZ	PAOZZ	PAOZZ	5310-00-809-4058	SCREW, ASSEMBLED .... ABC P-15121-78 (45722) 69-662-78 (30554)	EA 2	2	
						5305-01-078-		EA 4	4	
								EA 2	2	

(1) ILLUSTRATION (a) FIG NO.	(1) (b)	(2) SMR CODE a ARMY	b AIR FORCE	d NAVY	e USMC	(3) NATIONAL STOCK NUMBER	(4) DESCRIPTION / REF NUMBER MFR CODE — USABLE ON CODE	(5) U/M	(6) QTY INC IN	(7) USMC QTY PER EQUIP
							GROUP 02 FRAME			
5	12	XBOZZ	XBOZZ	XBOZZ	XBOZZ		BRACKET, GUARD.............................. ABC 134-4684 (44940) 84-13218 (30554)	EA	1	1
5	13	PAFZZ	PAOZZ	PAOZZ	PAOZZ	5320-00-165-8772	RIVET, SOLID .... ABC MS20426B4-6 (96906)	EA	4	33
5	14	PAFZZ	PAOZZ	PAOZZ	PAOZZ	5325-00-449-2967	RECEPTACLE.............................. ABC 99947P130 (61864)	EA	2	2
5	15	PAOZZ	PAOZZ	PAOZZ	PAOZZ	5975-00-878-3791	ROD, GROUND.................................. ABC 3598 (07464) FS0216B122-1 (15277)	EA	1	1
5	16	PAOZZ	PAOZZ	PAOZZ	PAOZZ	5340-01-275-3527	STRAP, RETANINNG.......................... ABC 403-2362 (44940) 84-13083 (30554)	EA	1	1
5	17	PAOZZ	PAOZZ	PAOZZ	PAOZZ	530640-226425	BOLT, MACHINE.................................. ABC MS9728-32 (96906)	EA	1	21
5	18	PAOZZ	PAOZZ	PAOZZ	PAOZZ	5310-00407-9566	WASHER, LOCK .... ABC MS35338-45 (96906) 850-1045 (44940)	EA	1	1
5	19	PAOZZ	PAOZZ	PAOZZ	PAOZZ	5310401-4219	WASHER, FLAT .... ABC MS27183-12 (96906)	EA	1	1
5	20	PAOZZ	PAOZZ	PAOZZ	PAOZZ	6145-0-189466	GROUND WIRE NO.6 .... ABC J-C-30AVA06CJ1/6AVAB0 (81348)	FT	5	5
5	21	PAOZZ	PAOZZ	PAOZZ	PAOZZ	59994)0-16-3912	CLAMP, ELECTRICAL.......................... ABC 70-801074 (04655)	EA	1	1
5	22	PAOZZ	PAOZZ	PAOZZ	PAOZZ	597540-9249927	DRIVE HEAD.. ABC GRB58 (73616)	EA	1	1
5	23	PAFZZ	PAOZ	PAOZZ	PAOZZ	532-01-049-8263	RIVET, BLIND....................................... ABC RV630-4-2 (53551) 810-0076 (44940) 72-5018 (30554)	EA	2	2
5	24	MDOZZ	MDOZZ	MDOZZ	MDOZZ	9905-01-120-8728	PLATE, IDENTIFICATION .... ABC 72-5029 (30554)	EA	1	1
5	25	PAOZZ	PAOZZ	PAOZZ	PAOZZ	5310-00-931-8167	NUT, PLAIN, HEXAGON .... ABC MS519676 (96906)	EA	6	6

(1) ILLUSTRATION (a) FIG NO.	(1) ILLUSTRATION (b)	(2) SMR CODE a ARMY	(2) SMR CODE b AIR FORCE	(2) SMR CODE d NAVY	(2) SMR CODE e USMC	(3) NATIONAL STOCK NUMBER	(4) DESCRIPTION REF NUMBER MFR CODE	(4) USABLE ON CODE	(5) U/M	(6) QTY INC IN	(7) USMC QTY PER EQUIP
							GROUP 02 FRAME				
5	26	PAOZZ	PAOZZ	PAOCZ	PAOZZ	5310-00-407-9566	WASHER, LOCK MS35338-45 (96906) 850-1045 (44940)	ABC	EA	6	6
5	27	PAOZZ	PAOZZ	PAOZZ	PAZZ	5310-00-081-4219	WASHER, FLAT MS27183-12 (96906)	ABC	EA	12	12
5	28	PAOZZ	PAOZZ	PAOZZ	PAOZZ	5306-00-226-4830	BOLT, MACHINE.................................. MS90728-37 (96906)	ABC	EA	6	6
5	29	XBF2Z	XBFZZ	XBFZZ	XBFZZ		FRAME... 403-2364 (44940) 84-1320 (30554)	ABC	EA 1	1	

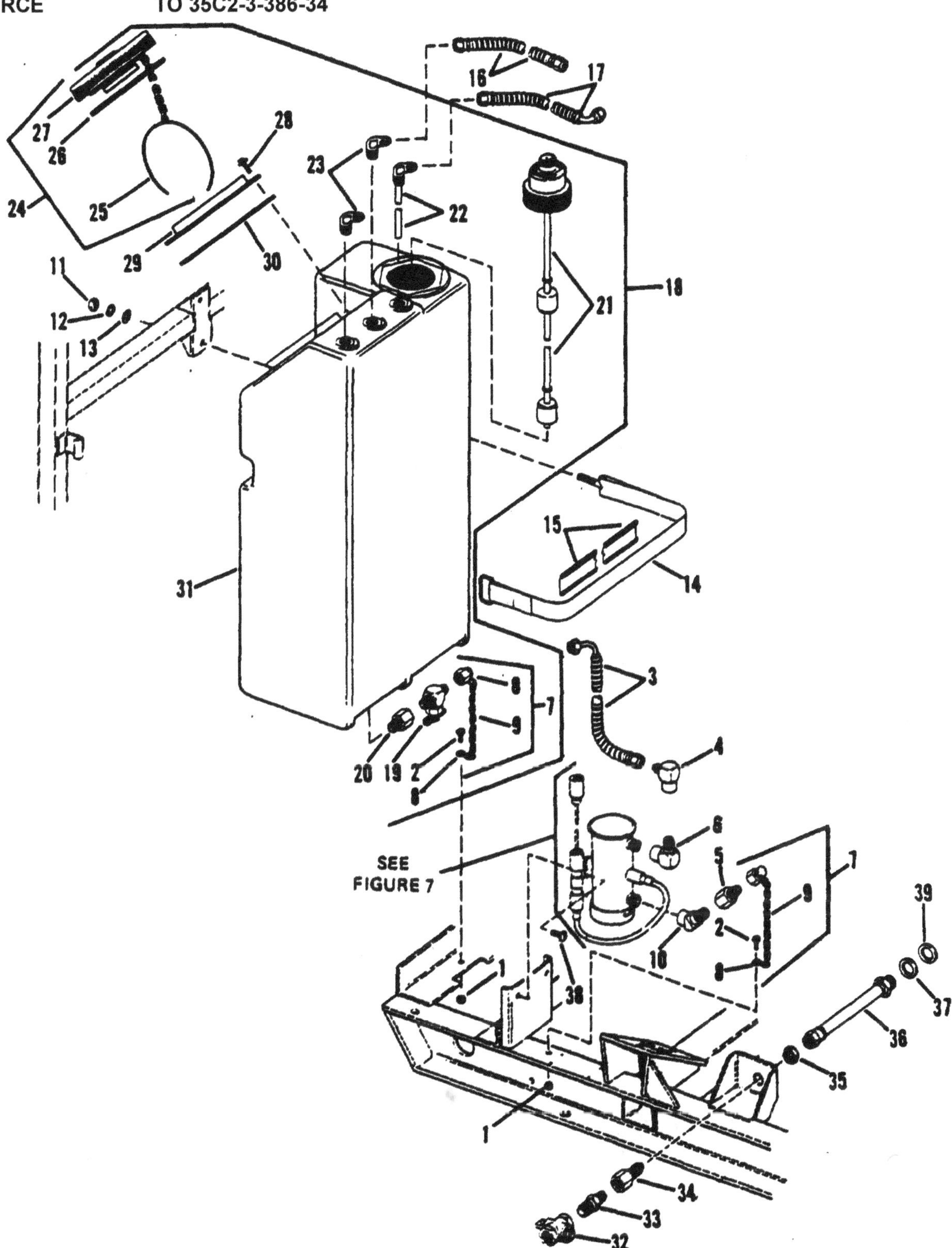

Figure 6. Fuel Tank, Lines and Oil Drain, Group 04 (Fuel System and Oil Drain)

(1) ILLUSTRATION (a) FIG NO.	(1) ILLUSTRATION (b)	(2) SMR CODE a ARMY	(2) SMR CODE b AIR FORCE	(2) SMR CODE d NAVY	(2) SMR CODE e USMC	(3) NATIONAL STOCK NUMBER	(4) DESCRIPTION REF NUMBER MFR CODE	(4) USABLE ON CODE	(5) U/M	(6) QTY INC IN	(7) USMC QTY PER EQUIP
							GROUP 04 FUEL SYSTEM AND OIL DRAIN				
6	1	PAOZZ	PAOZZ	PAOZZ	PAOZZ	5310-0596-173	NUT, PLAIN, ASSEMBLED 501-250800-00 (78189) 69-561-5 (30554)	ABC	EA	2	2
6	2	PAOZZ	PAOZZ	PAOZZ	PAOZZ	5305-01-078-5064	SCREW, ASSEMBLED.......................... P-15121-78 (45722) 69-662-78 (30554)	ABC	EA	2	2
6	3	PAOZZ	PAOZZ	PAOZZ	PAOZZ	4720-00-275-6168	HOSE ASSY, NONNETALLIC FL3650EEE-0187 (01276) 84-13008 (30554)	ABC	EA	2	2
6	4	PAOZZ	PAOZZ	PAOZZ	PAOZZ		ELBOW, 90 DEG 4-4070203BA (81343) 4-4BTX-B (98441)	ABC	EA	1	1
6	5	PAOZZ	PAOZZ	PAOZZ	PAOZZ	4730-00-277-7939	ADAPTER, STRAIGHT.......................... 5-4GBTX (98441)	ABC	EA	1	1
6	6	PAOZZ	PAOZZ	PAOZZ	PAOZZ	4730-00-24-0497	ELBOW, PIPE 1-4X1-4CDB (98441)	ABC	EA	1	1
6	7	PAOZZ	PAOZZ	PAOZ	PAOZZ	4730-01-103-1199	CAP ASSY... SP272FM (70411) 502-0318 (44940) 69-539-2 (30554)	ABC	EA	2	2
6	8	XAOZZ	XAOZZ	XAOZZ	XAOZZ	4730-00-812-1333	HOOK MS87006-3 (96906)	ABC	EA	4	4
6	9	XAOZZ	XAOZZ	XAOZZ	XAOZZ		CHAIN, WELDLESS RRC271 (81348)	ABC	EA	2	2
6	10	PAOZZ	PAOZZ	PAOZZ	PAOZZ	4030-00-270-5436	ELBOW, 45 DEG.................................. X121619 (80072) 4-4 130339B (81343)	ABC	EA	1	1
6	11	PAOZZ	PAOZZ	PAOZZ	PAOZZ	4010-00-809-2719	NUT, PLAIN, HEXAGON MS51968-6 (9690)	ABC	EA	1	1
6	12	PAOZZ	PAO=	PAOZZ	PAOUZ	4730-00-278-4497	WASHER, LOCK MS3533-45 (96906) 850-1045 (44940)	ABC	EA	1	1
6	13	PAOZZ	PAOZZ	PAOZZ	PAOZZ	5310-00-905-4600	WASHER, FLAT MS27183-12 (96906)	ABC	EA	1	1
6	14	PAFZZ	PAFZZ	PAFZZ	PAFZZ	5310-00-407-9566	BAND, RETAINING 159-1135 (44940) 84-13005 (30554)	ABC	EA	1	1
						5310-00-081-					

(1) ILLUSTRATION (a) FIG NO.	(b)	(2) SMR CODE a ARMY	b AIR FORCE	d NAVY	e USMC	(3) NATIONAL STOCK NUMBER	(4) DESCRIPTION REF NUMBER MFR CODE	USABLE ON CODE	(5) U/M	(6) QTY INC IN	(7) USMC QTY PER EQUIP
							GROUP 04 FUEL SYSTEM AND OIL DRAIN				
6	15	PAFZZ	PAFZZ	PAFZZ	PAFZZ	5340-01-275-3404	BRACKET, DOUBLE ANGLE 159-1138 (44940) 84-13006 (30554)	ABC	EA	1	1
6	16	PAFZZ	PAFZZ	PAFZZ	PAFZZ	4720-01-275-2441	HOSE ASSY, NONMETALLIC............... FL5958EE-0263 (01276) 501-0506 (44940) 84-13011 (3054)	ABC	EA	1	1
6	17	PAFZZ	PAFZ	PAFZZ	PAFZZ	4720-01-275-c1C9	HOSE ASSY, NONMETALLIC............... FL3650EEE0307 (01276) 501-504 (44940) 84-13009 (30554)	ABC	EA	2	2
6	18	AFFFF	AFFFF	AFFFF	AFFFF		TANK ASSY, FUEL 159-1143 (44940) 84-13073 (30554)	ABC	EA	1	1
6	19	PAOZZ	PAOZZ	PAOZZ	PAOZZ		COCK, SHUT-OFF........................ 695 (79470) 72-5101 (30554)	ABC	EA	1	1
6	20	PAOZZ	PAOZZ	PAOZZ	PAOZZ	420-00-595-3191	COUPLING, PIPE.......................... 1/4FG-B (98441) 4-4 140139B (81343)	ABC	EA	1	1
6	21	PAOZ	PAOZZ	PAOZZ	PAOZZ		SWITCH, LIQUID LEVEL.............. 120012 (04034) 84-13001 (30554)	ABC	EA	1	1
6	22	PAOZZ	PAOZZ	PAOZZ	PAOZZ	5930-01-260-5443	TUBE ASSY, METAL............................ 159-1142 (44940) 84-1302 0(30554)	ABC	EA	1	1
6	23	PAOZZ	PAOZZ	PAOZZ	PAOZZ	4710-01-274-492-	ELBOW, PIPE TO TUBE 4-4070202BA (81343) 4-4CBTXB (98441)	ABC	EA	2	2
6	24		AOOOO			4730-00-540-1861	CAP, FILLER OPENING....................... 159-1137 (44940) 84-13050 (30554)	ABC	EA	1	1
6	25		PAOZZ			5340-01-305-3401	RETAINER, CAP 159-1136 (44940) 84-13051 (30554)	ABC	EA	1	1
6	26		PAOZZ				GASKET MS35643-1 (96906)	ABC	EA	1	1
6	27		PAOZZ			2915-01-294-1815	CAP, FUELTANK.......................... MS35645-2 (96906)	ABC	EA	1	1
						5330684-78S51					

(1) ILLUSTRATION (a) FIG NO.	(1) ILLUSTRATION (b)	(2) SMR CODE a ARMY	(2) SMR CODE b AIR FORCE	(2) SMR CODE d NAVY	(2) SMR CODE e USMC	(3) NATIONAL STOCK NUMBER	(4) DESCRIPTION REF NUMBER MFR CODE	USABLE ON CODE	(5) U/M	(6) QTY INC IN	(7) USMC QTY PER EQUIP
							GROUP 04 FUEL SYSTEM AND OIL DRAIN				
6	28	PAOZZ	PAOZZ	PAOZZ	PAOZZ	5305-00-036-6906	SCREW, ASSEMBLED P-15121-50 (45722) 69662-50 0PS4)	ABC	EA	8	13
6	29	XBOZZ	XBOZZ	XBOZZ	XBOZZ		FLANGE, FILLER NECK 159-1175 (44940) 84-13286 (30554)	ABC	EA	1	1
6	30	PAOZZ	PAOZZ	PAOZZ	PAOZZ	5330-01-275-3368	GASKET 159-1176 (44940) 4-1327 (3054)	ABC	EA	1	1
6	31	PAFZZ	PAFZZ	PAFZZ	PAFZZ	2910-01-275-l749	TANK, FUEL 84-13000 (1DS87) 84-13000 (30554)	ABC	EA	1	1
6	32	PAOZZ	PAOZZ	PAOZZ	PAOZZ	4820-00-136-1085	COCK, OIL DRAIN................................. 1260-1-2 (9'000) 132l6E3329-1 (97403)	ABC	EA	1	1
6	33	PAOZZ	PAOZZ	PAOZZ	PAOZZ		REDUCER, PIPE................................... 3/S8Xl/4FFB (98441) J514 (81343)	ABC	EA	1	1
6	34	PAOZZ	PAOZZ	PAOZZ	PAOZZ	4730-01-277-1399	ADAPTER, STRAIGHT 6WGBTXS (98441)	ABC	EA	1	1
6	35	PAOZZ	PAOZZ	PAOZZ	PAOZZ	4730-01-050-3941	NUT, PLAIN, HEXAGON 2P1279 (11083) 6WLNS (98441)	ABC	EA	1	1
6	36	PAOZZ	PAOZZ	PAOZZ	PAOZZ	5310-00-186-9550	HOSE ASSY, NONMETALLIC FB9876-01-0070 (01276) 84-13098 (30554)	ABC	EA	1	1
6	37	PAOZZ	PAOZZ	PAOZZ	PAOZZ	4720-01-275-5225	NUT, PLAIN, ROUND 102-1298 (4940) U-13205 (0554)	ABC	EA	1	1
6	38	PAOZZ	PAOZZ	PAOZZ	PAOZZ		SCREW, ASSEMBLED P-15121-78 (45722) 69-662.78 (30554)	ABC	EA	2	2
6	39	PAOZZ	PAOZZ	PAOZZ	PAOZZ	5310-01-275-3307	GASKET.. 102-1147 (44940) C0102114700 (15434)	ABC	EA	1	1
						5305-01-078-5064					

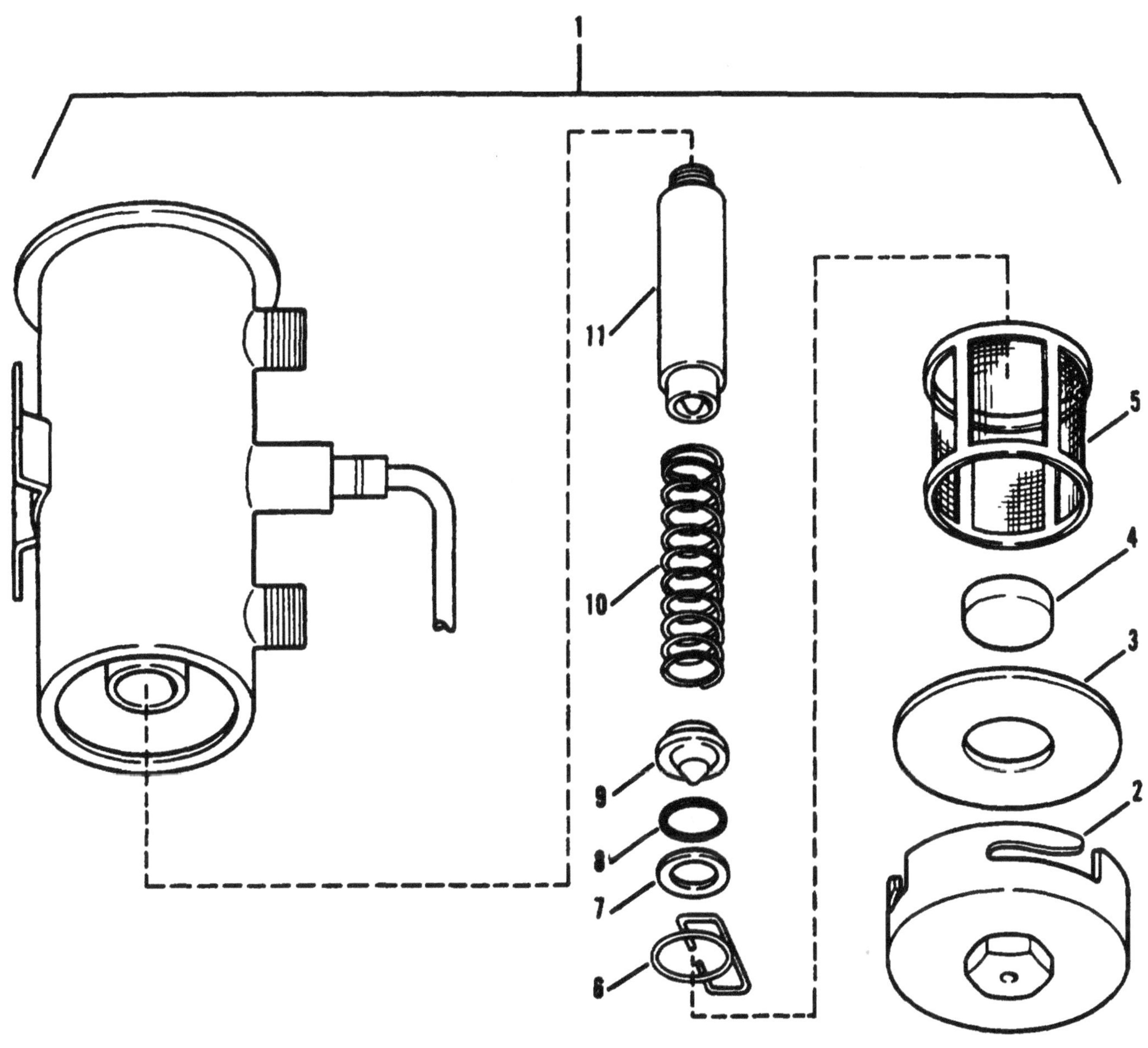

Figure 7. Auxiliary Fuel Pump, Group 04 (Fuel System and Oil Drain)

(1) ILLUSTRATION (a) FIG NO.	(b)	(2) SMR CODE a ARMY	b AIR FORCE	d NAVY	e USMC	(3) NATIONAL STOCK NUMBER	(4) DESCRIPTION REF NUMBER MFR CODE	USABLE ON CODE	(5) U/M	(6) QTY INC IN	(7) USMC QTY PER EQUIP
							GROUP 04 FUEL SYSTEM AND OIL DRAIN				
7	1	PAOZZ	PAOZZ	PAOZZ	PAOZZ	2910-01-276-1484	PUMP, FUEL, ELECTRICAL 40128 (72850) 84-13139 (30554)	ABC	EA	1	1
7	2	PAOZZ	PAOZZ	PAOZZ	PAOZZ	2910-01-063-3144	COVER ASSY 479130 (72850)	ABC	EA	1	1
7	3	PAOZZ	PAOZZ	PAOZZ	PAOZZ	5330-00-763-9322	GASKET 479136 (72850)	ABC	EA	1	1
7	4	PAOZZ	PAOZZ	PAOZZ	PAOZZ	4730-00-483-5176	PLUG, MAGNETIC 479012 (72850)	ABC	EA	1	1
7	5	PAOZZ	PAOZZ	PAOZZ	PAOZZ	2910-00-893-6402	STRAINER ELEMENT 479729 (72850)	ABC	EA	1	1
7	6	PAOZZ	PAOZZ	PAOZZ	PAOZZ	4730-01-095-5584	CLAMP, HOSE 479137 (72850)	ABC	EA	1	1
7	7	PAOZZ	PAOZZ	PAOZZ	PAOZZ	5310-00-199-1789	WASHER, FLAT 479138 (72850)	ABC	EA	1	1
7	8	PAOZZ	PAOZZ	PAOZZ	PAOZZ	5330-00-220-2631	PACKING, PREFORMED 479139 (72850)	ABC	EA	1	1
7	9	PAOZZ	PAOGZ	PAOZZ	PAOZZ	5360-01-094-2717	CUP ASSEMBLY, SPRING 479132 (72850)	ABC	EA	1	1
7	10	PAOZZ	PAOOZ	PAOZZ	PAOZZ	5360-00-200-9691	SPRING, HELICAL 476285 (72850) 10947657 (19207)	ABC	EA	1	1
7	11	PAOZZ	PAOZZ	PAOZZ	PAOZZ		PLUNGER 41625 (72850)	ABC	EA	1	1

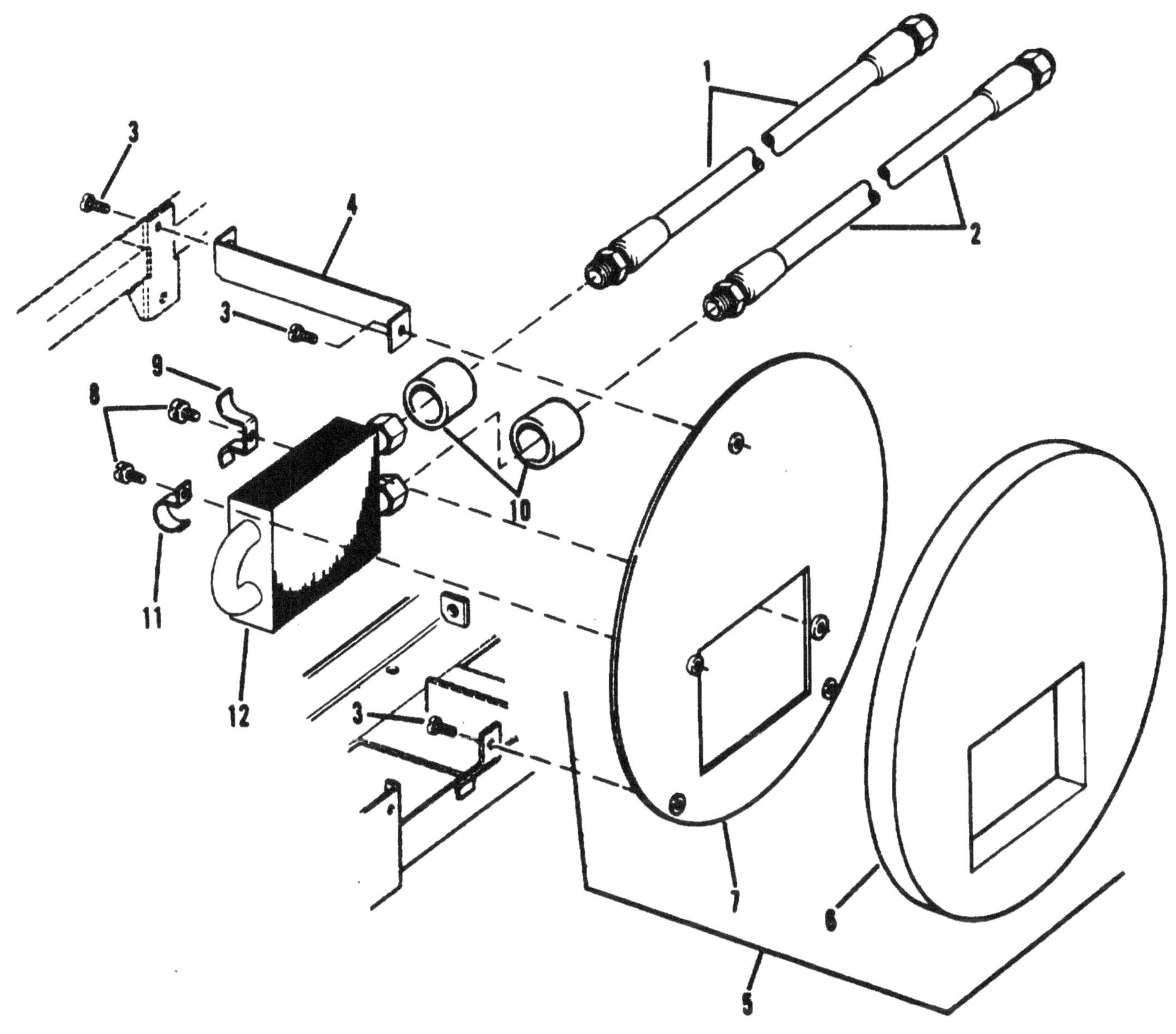

Figure 8. Oil Cooler, Group 05 (Engine)

(1) ILLUS-TRA-TION (a) FIG NO.	(b)	(2) SMR CODE a ARMY	b AIR FORCE	d NAVY	e USMC	(3) NATIONAL STOCK NUMBER	(4) DESCRIPTION USABLE ON CODE REF NUMBER MFR CODE	(5) U/M	(6) QTY INC IN	(7) USMC QTY PER EQUIP
							GROUP 05 ENGINE			
8	1	PAFZZ	PAFZZ	PAFZZ	PAFZZ	4720-01-274-4749	HOSE ASSY, NONMETALLIC .... ABC 503-1517-01 (44940)	EA 1	1	
8	2	PAFZZ	PAFZZ	PAFZZ	PAFZZ	4720-01-274-4749	HOSE ASSY, NONMETALLIC .... ABC 503-1517-01 (44940)	EA 1	1	
8	3	PAFZZ	PAFZZ	PAFZZ	PAFZZ	5305-01-078-5064	SCREW, ASSEMBLED .... ABC P-15121-78 (45722) 69-662-78 (30554)	EA 5	5	
8	4	XBFZZ	XBFZZ	XBFZZ	XBFZZ		SUPPORT,AIR BAFFLE .... ABC 134-4571 (44940) 84-13142 (30554)	EA 1	1	
8	5	XBFZZ	XBFZZ	XBFZZ	XBFZZ		BAFFLE ASSY .... ABC 134-4570 (44940) 84-13109 (30554)	EA 1	1	
8	6	PAFZZ	PAFZZ	PAFZZ	PAFZZ		INSULATION .... ABC 134-4569 (44940) 84-13308-2 (30554)	EA 1	1	
8	7	XBFZZ	XBFZZ	XBFZZ	XBFZZ	5640-01-207-2042	PLATE, AIR BAFFLE .... ABC 134-4568 (44940) 84-13111 (30554)	EA 1	1	
8	8	PAFZZ	PAPAFZ	PAFZZ	PAFZZ		SCREW, ASSEMBLED .... ABC P15121-64 (45722) 69-662-64 (30554)	EA 1	1	
8	9	PAFZZ	PAFZZ	PAFZZ	PAFZZ	5305-00-1916226	STRAP, RETAINING .... ABC 102-1344 (44940) 84-13278 (30554)	RA 1	1	
8	10	PAFZZ	PAFZZ	PAFZZ	PAFZZ	5340-01-275-3529	BUSHING, NONMETALLIC .... ABC TA03M72CR12 (84971) 84-13338 (30554)	EA	2	2
8	11	PAFZZ	PAFZZ	PAFZZ	PAFZZ	5365-01-277-4616	CLAMP, LOOP .... ABC MS21333-76 (96906)	EA 1	1	
8	12	PAFZZ	PAFZZ	PAFZZ	PAFZZ	5340--724-7038	COOLER, FLUID .... ABC 102-1321-01 (44940) 017536 (50184)	EA	1	1

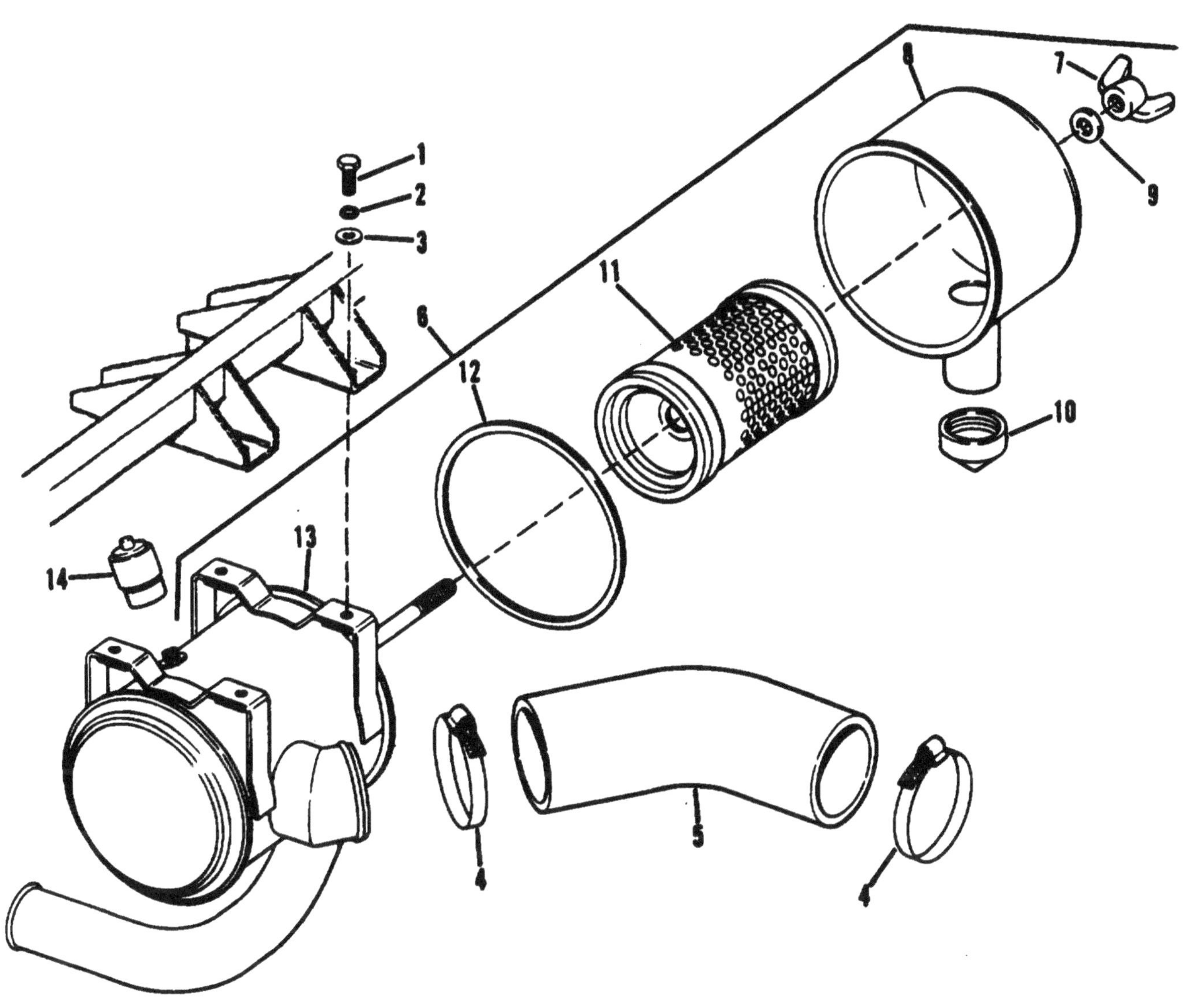

Figure 9. Air Intake Filter, Group 05 (Engine)

(1) ILLUSTRATION (a) FIG NO.	(b)	(2) SMR CODE a ARMY	b AIR FORCE	d NAVY	e USMC	(3) NATIONAL STOCK NUMBER	(4) DESCRIPTION REF NUMBER MFR CODE	USABLE ON CODE	(5) U/M	(6) QTY INC IN	(7) USMC QTY PER EQUIP
							GROUP 05 ENGINE				
9	1	PAOZZ	PAOZZ	PAOZZ	PAOZZ	5306-00-226-4825	BOLT, MACHINE MS90728-32 (96906)	ABC	EA 4	4	
9	2	PAOZZ	PAOZZ	PAOZZ	PAOZZ	5310-00-407-9566	WASHER, LOCK MS35338-45 (96906) 850-1045 (44940)	ABC	EA	4	4
9	3	PAOZZ	PAOZZ	PAOZZ	PAOZZ		WASHER, FLAT MS27183-12 (96906)	ABC	EA	4	4
9	4	PAOZZ	PAOZZ	PAOZZ	PAOZZ	5310-00-081-4219	CLAMP,HOSE MS35842-14 (96906)	ABC	EA 2	2	
9	5	PAOUZ	PAO=	PAOZZ	PAOZZ	4730-00 908-6292	HOSE, PREFORMED 503-1464 (44940) 84-13291 (30554)	ABC	EA 1	1	
9	6	PAOZZ	PAOZZ	PAOZZ	PAOZZ	4720-01-274-4933	AIR CLEANER, INTAKE 75509-N (76700) 140-1990 (44940) 84-13290 (30554)	ABC	EA 1	1	
9	7	PAOZZ	PAOZZ	PAOZZ	PAOZZ	2940-01-274-6800	NUT, SELF-LOCKING Q01444 (76700) 84-13328 (30554)	ABC			
9	8	XAOZZ	XAOZZ	XAOZZ	XAOZZ	5310-01-277-7334	COVER, FILTER Q33639 (76700) 4-13329 (30554)	ABC	EA	1	1
9	9	PAOZZ	PAOZZ	PAOZZ	PAOZZ		GASKET Q66993 (76700) 84-13335 (30554)	ABC	EA 1	1	
9	10	PAOZZ	PAOZZ	PAO2Z	PAOZZ		VALVE, FLAPPER Q59400 (76700) 84-13330 (30554)	ABC	EA 1	1	
9	11	PAOZZ	PAOZZ	PAOZZ	PAOZZ	5330-01-280-9392	ELEMENT, FILTER 84-13336 (30554) CS-4310-SV-0728 (16236)	ABC	EA 1	1	
9	12	PAOZZ	PAOZZ	PAOZZ	PAOZZ	4820-01-282-0165	PACKING, PREFORMED Q04190 (76700) 84-13337 (30554)	ABC	EA 1	1	
9	13	XAOZZ	XAOZZ	XAOZZ	XAOZZ	4310-01-281-5988	BODY, FILTER 84-13331 (30554)	ABC			
9	14	PAOZZ	PAOZZ	PAOZZ	PAOZZ	5330-01-287-0922	INDICATOR, PRESSURE RBX00-2280 (18265) 140-0961 (44940) 69-788 (30554)	ABC	EA 1	1	
									EA 1	1	
									EA 1	1	

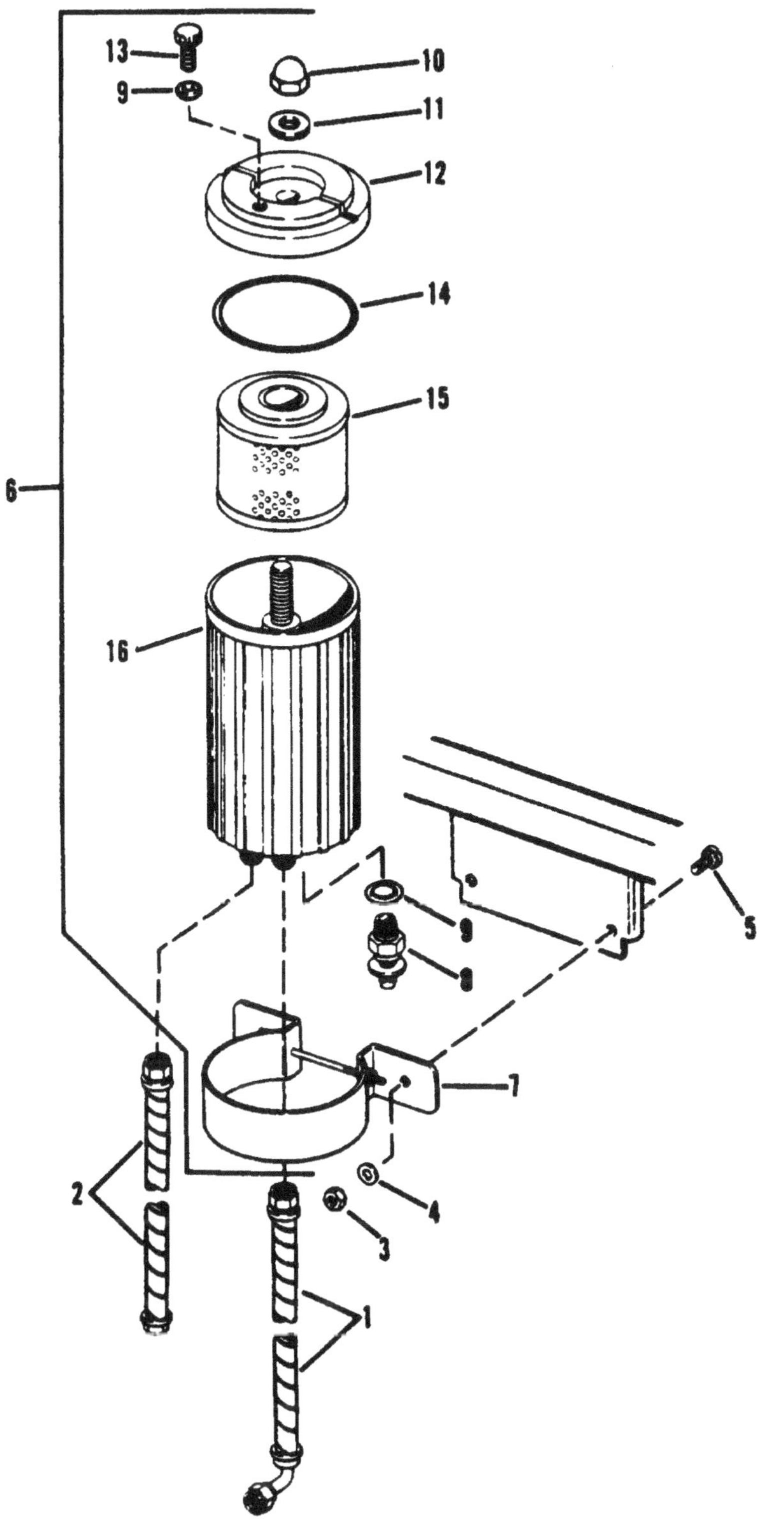

Figure 10. Fuel Filter/Water Separator, Group 04 (Fuel System and Oil Drain)

(1) ILLUSTRATION (a) FIG NO.	(b)	(2) SMR CODE a ARMY	b AIR FORCE	d NAVY	e USMC	(3) NATIONAL STOCK NUMBER	(4) DESCRIPTION USABLE ON CODE REF NUMBER MFR CODE	(5) U/M	(6) QTY INC IN	(7) USMC QTY PER EQUIP
							GROUP 04 FUEL SYSTEM AND OIL DRAIN			
10	1	PAOZZ	PAOZZ	PAOZZ	PAOZZ	4720-01-288-0786	HOSE ASSY, NONMETALLIC............... ABC FL5114EEE-0091 (01276) 84-13010 (30554!)	EA 1	1	
10	2	PAOZ	PAOZZ	PAOZZ	PAOZZ	4720-01-281-5193	HOSE ASSY, NONMETALLIC............... ABC FA1796EEE-0111 (01276) 84-13334 (30554)	EA 1	1	
10	3	PAOZZ	PAOZZ	PAOZZ	PAOZZ	5310-00-696-5173	NUT, PLAIN, ASSEMBLED .... ABC 501-250800-00 (78189) 69-561-5 (30554)	EA	2	2
10	4	PAOZZ	PAOZZ	PAOZZ	PAOZZ		WASHER, FLAT .... ABC MS271I3-10 (96906)	EA	2	2
10	5	PAOZZ	PAOZZ	PAOUE	PAOZZ	5310-00-809-4058	SCREW, ASSEMBLED......................... ABC 69-662-78 (30554) P-15121-78 (45722)	EA 2	2	
10	6	PAOZZ	PAOZZ	PAOZZ	PAOZZ	5305-01-078-5064	FILTER, FLUID .... ABC 192100 (0A569) 84-13208 (30554)	EA	1	1
10		PAOZZ	PAOZZ	PAOZZ	PAOZZ	2910-01-275-8028	PARTS KIT, FLUID .... ABC 192050 (0A569) 84-13324 (30554)	EA 2	1	
10	11	PAOZZ	PAOZZ	PAOZZ	PAOZZ	2910-01-275-2460	PACKING, PREFORMED ...... ABC 03(212 (0A569) 5o9-S237 (44940)	A 1	1	
10	9	PAOZZ	PAOZZ	PAOZZ	PAOZZ	5330-01-278-9475	PACKING, PREFORMED ...... ABC 031523 (0A569) 5094236 (44940)	EA 2	2	
10	14	PAOZZ	PAOZZ	PAOZZ	PAOZZ	5330-01-283-2407	GASKET .... ABC 101005 (0A569) 50938 (44940)	EA 1	1	
10	15					5330-01-276-6707	ELEMENT, FILTER................ ABC PART OF KIT 192050 (OA569)	EA 1	1	
10	7	XBOZZ	XBOZZ	XBOZZ	XBOZZ		BRACKET, FILTER....................... ABC 151051 (OA569) 8413327 (30554)	EA 1	1	
10	8	PAOZZ	PAOZZ	PAOZZ	PAOZZ		COCK, DRAIN, FUEL .... ABC 153022 (0A569) 84-13326 (30554)	EA 1	1	
								EA 1	1	

(1) ILLUSTRATION (a) FIG NO.	(b)	(2) SMR CODE a ARMY	b AIR FORCE	d NAVY	e USMC	(3) NATIONAL STOCK NUMBER	(4) DESCRIPTION REF NUMBER MFR CODE	USABLE ON CODE	(5) U/M	(6) QTY INC IN	(7) USMC QTY PER EQUIP
							GROUP 04 FUEL SYSTEM AND OIL DRAIN				
10	10	PAOZZ	PAOZZ	PAOZZ	PAOZZ	5310-01-278-8506	NUT, PLAIN, CAP 192001 (0A569) 869-0006 (44940)	ABC	EA	1	1
									EA	1	2
10	12	XAOZZ	XAOZZ	XAOZZ	XAOZZ		COVER, FILTER 192012 (0A569) 84-13322 (30554)	ABC	EA	1	1
10	13	PAOZZ	PAOZZ	PAOZZ	PAOZZ	5305-01-276-1629	THUMBSCREW 191081 (0A569) 84-13320 (30554)	ABC	EA	1	1
									EA	1	2
									EA	1	2
10	16	XAOZZ	XAOZZ	XAOZZ	XAOZZ		BODY, FILTER 149-2147 (44940)	ABC	EA	1	1

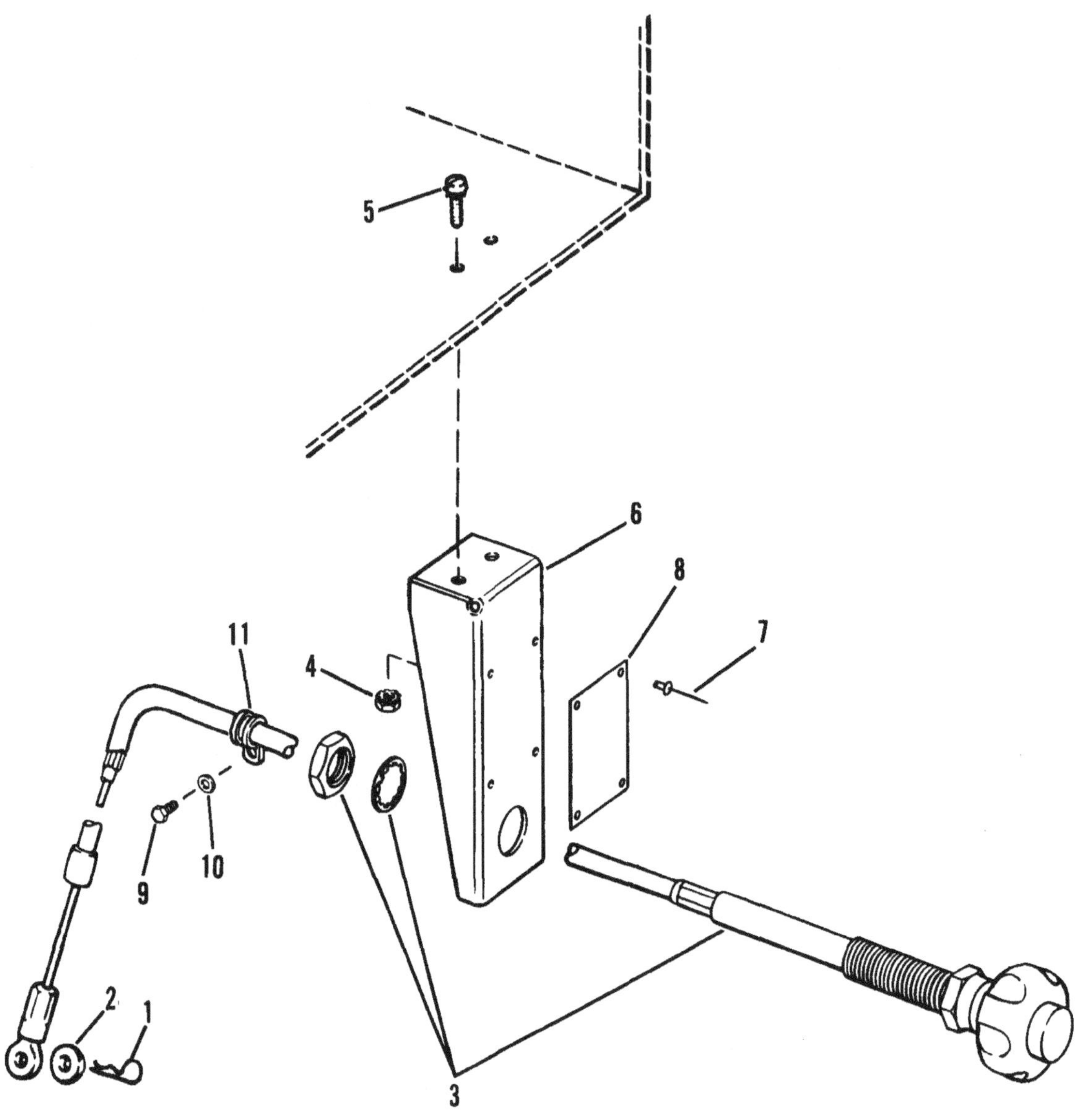

Figure 11. Governor Speed Control, Group 07 (Engine Control)

(1) ILLUS-TRA-TION (a) FIG NO.	(b)	(2) SMR CODE a ARMY	b AIR FORCE	d NAVY	e USMC	(3) NATIONAL STOCK NUMBER	(4) DESCRIPTION USABLE ON CODE REF NUMBER MFR CODE	(5) U/M	(6) QTY INC IN	(7) USMC QTY PER EQUIP
GROUP 07 ENGINE CONTROLS										
11	1	PAOZZ	PAOZZ	PAOZZ	PAOZZ	5315-00-839-5820	PIN, COTTER........ ABC MS24665-134 (96906)	EA 1	1	
11	2	PAOZZ	PAOZZ	PAOZZ	PAOZZ	5310-01-276-3342	WASHER, FLAT .... ABC 526-0016 (44940)	EA	1	1
11	3	PAOZZ	PAOZZ	PAOZZ	PAOZZ	3040-01-274-6809	CONTROL ASSY, PUSH-PULL........ ABC 345-029-27 (70436) 84-13074 (30554)	EA 1	1	
11	4	PAOZZ	PAOZZ	PAOZZ	PAOZZ	5310-00-696-5173	NUT, PLAIN, ASSEMBLED .... ABC 501-250800-00 (78189) 69-561-5 (30554)	EA	2	2
11	5	PAOZZ	PAOZZ	PAOZZ	PAOZZ		SCREW, ASSEMBLED........ ABC P-15121-78 (45722) 69-662-78 (30554)	EA 2	2	
11 XBOZZ	6	XBOZ		XBOZZ	XBOZZ	5305-01-078-5064	BRACKET.SPEED CONTR .... ABC 150-2109 (44940) 84-13081 (30554)	EA 1	1	
11	7	PAFZZ	PAOZZ	PAOZZ	PAOZZ		RIVET, BLIND........ ABC 818-0076 (44940) 72-5018 (30554) RV630-4-2 (53551)	EA 4	4	
11	8	MDFZZ	MDOZZ	MDOZZ	MDOZZ	5320-01-049-8263	PLATE, INSTRUCTION .... ABC 099-2321 (44940) 84-1302 (30554)	EA 1	1	
11	9	PAOZZ	PAOZZ	PAOZZ	PAOZZ		BOLT, MACHINE........ ABC 718-1018 (44940)	EA	1 35	
11	10	PAOZZ	PAOZZ	PAOZZ	PAOZZ	5306-01-275-6000	WASHER, FLAT .... ABC 740-1004 (4940) C0740100400 (15434)	EA	1 54	
11	11	PAOZZ	PAOZZ	PAOZZ	PAOZZ	5310-01-275-3318	CLAMP, LOOP .... ABC MS21333-102 (9696)	EA	1	1

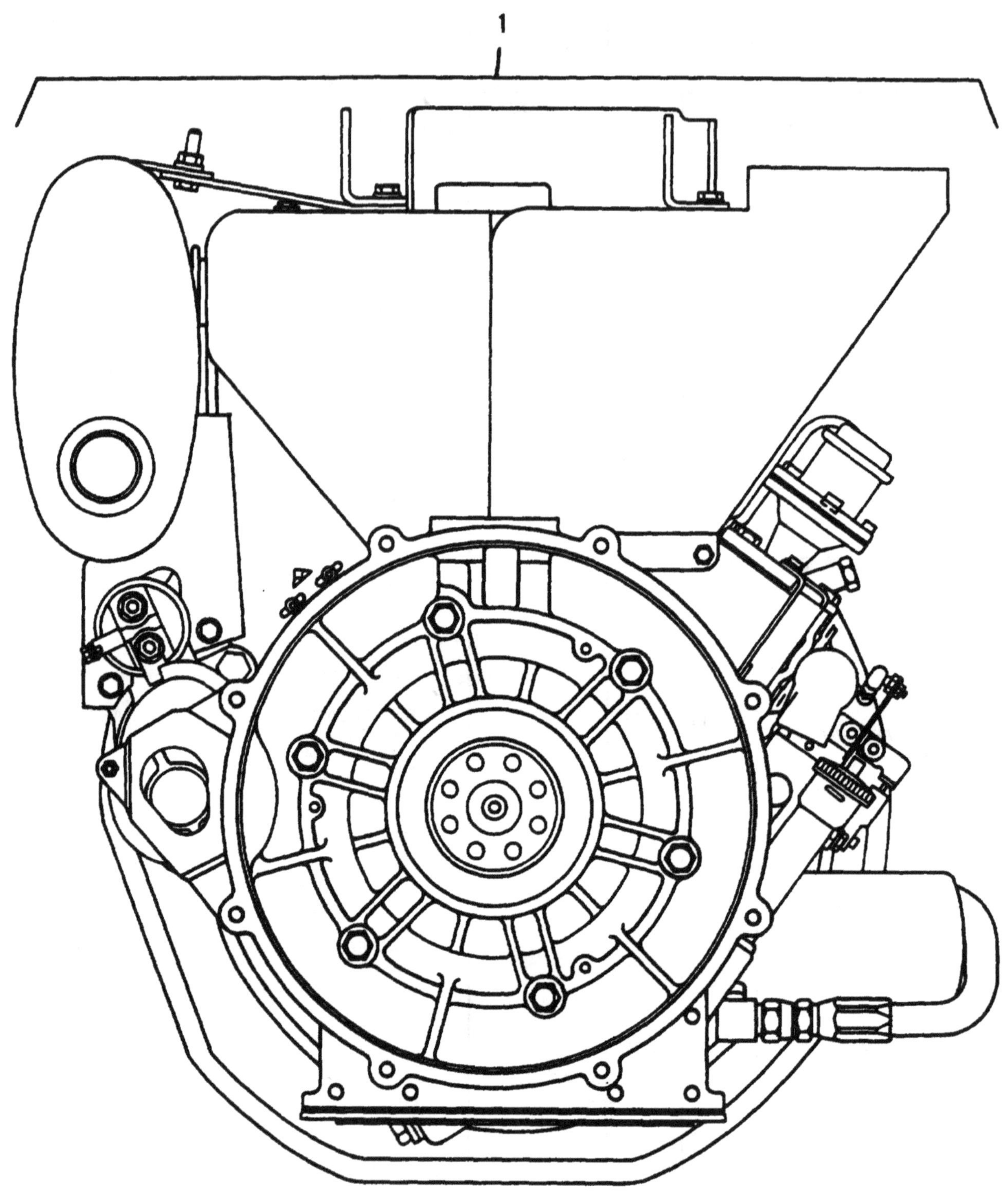

Figure 12. 3KW Diesel Engine, Group 05 (Engine)

(1) ILLUSTRATION (a) FIG NO.	(1) ILLUSTRATION (b)	(2) SMR CODE a ARMY	(2) SMR CODE b AIR FORCE	(2) SMR CODE d NAVY	(2) SMR CODE e USMC	(3) NATIONAL STOCK NUMBER	(4) DESCRIPTION USABLE ON CODE / REF NUMBER MFR CODE	(5) U/M	(6) QTY INC IN	(7) USMC QTY PER EQUIP
							GROUP 05 ENGINE			
12		PAHZZ	PAFZZ	PAFZZ	PAFZZ	5330-01-283-4297	GASKET AND PREFORMED .... ABC 168-0184 (44940)	EA 2	1	
12	1	XBFHH	XBFHH	XBFHB	XBFHH	2815-01-274-6803	ENGINE, DIESEL, 3KW .... ABC Q106DL10399 (44940) 84-13209 (30554)	EA 1	1	

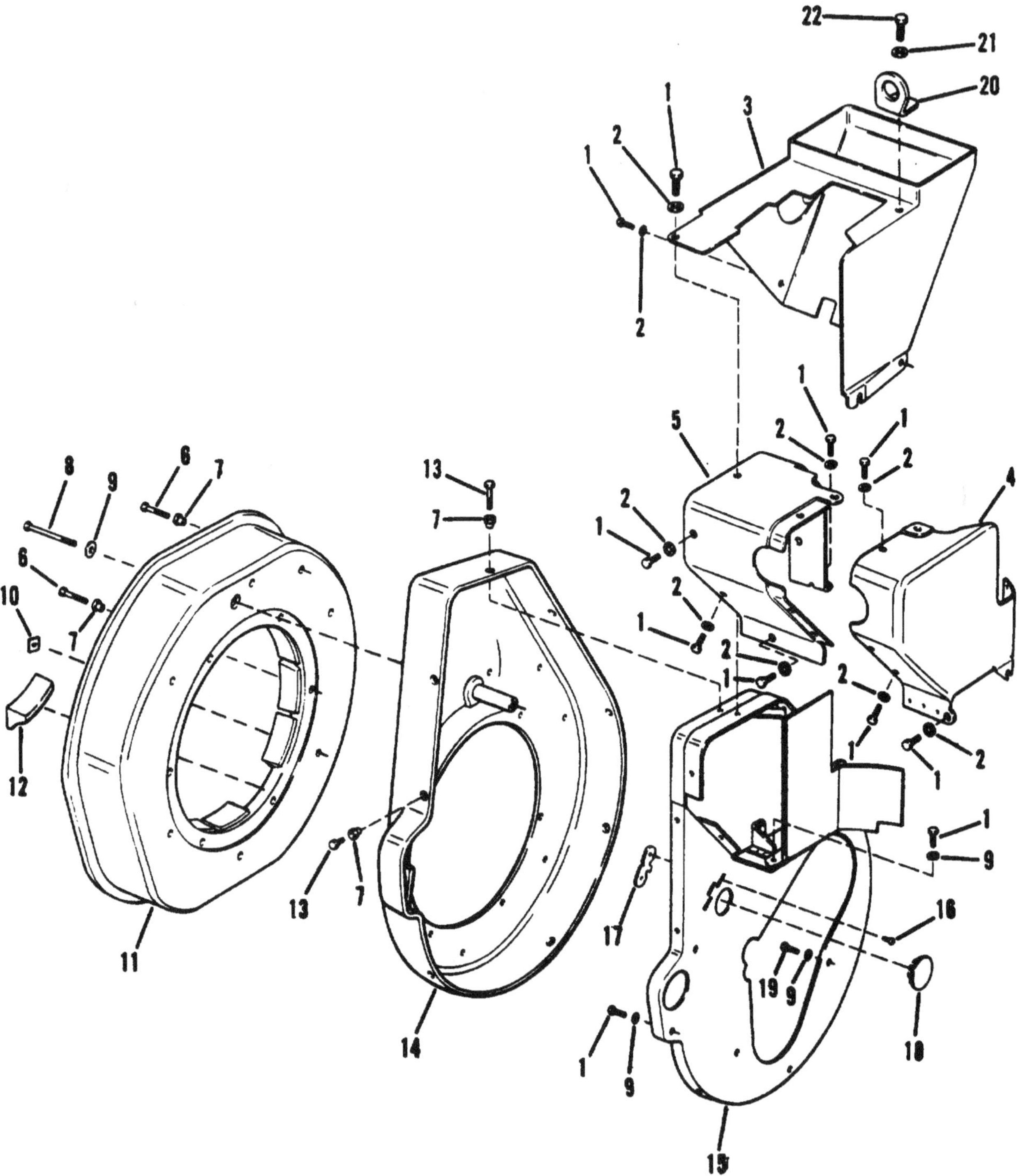

Figure 13. Engine Covers and Air Scroll, Group 05 (Engine)

(1) ILLUSTRATION (a) FIG NO.	(1) ILLUSTRATION (b)	(2) SMR CODE a ARMY	(2) SMR CODE b AIR FORCE	(2) SMR CODE d NAVY	(2) SMR CODE e USMC	(3) (4) NATIONAL STOCK NUMBER / DESCRIPTION USABLE ON CODE / REF NUMBER MFR CODE	(5) U/M	(6) QTY INC IN	(7) USMC QTY PER EQUIP
						GROUP 05 ENGINE			
13	1	PAFZZ	PAFZZ	PAFZZ	PAFZZ	5306-01-275-0000 BOLT, MACHINE .... ABC 718-1018 (44940)	EA	14	14
13	2	PAFZZ	PAFZZ	PAFZZ	PAFZ	5310-01-275-3316 WASHER, FLAT .... ABC 740-1004 (44940)	EA	12	12
13	3	XBO	XBOZZ	XBOZZ	XBOZZ	DUCT ASSY.OUTLET .... ABC 134-4606-01 (44940)	EA 1	1	
13	4	XBFZZ	XBFZZ	XBFZZ	XBFZZ	WRAPPER, REAR .... ABC 134-4573-01 (44940)	EA 1	1	
13	5	XBFZZ	XBFZZ	XBFZZ	XBFZZ	WRAPPER ASSY, FRONT .... ABC 134-4533-01 (44940)	EA 1	1	
13	6	PAFZZ	PAFZZ	PAFZZ	PAFZZ	5306-01-275-3267 BOLT, SELF-LOCKING .... ABC 815-0673 (44940)	EA 13	13	
13	7	PAFZZ	PAFZZ	PAFZZ	PAFZZ	3120-01-290-6902 BUSHING, SLEEVE .... ABC 134-4594 (44940)	EA	21	21
13	8	PAFZZ	PAFZZ	PAFZZ	PAFZZ	5305-01-276-0850 SCREW, SELF LOCKING .... ABC 815-0674 (44940)	EA 1	1	
13	9	PAFZZ	PAFZZ	PAFZZ	PAFZZ	5310-01-275-7786 WASHER, FLAT .... ABC 526-0018 (44940)	EA 9	9	
13	10	PAFZZ	PAFZZ	PAFZZ	PAFZZ	5310-01-276-1653 PUSH ON NUT .... ABC 870-0440 (44940)	EA	12	12
13	11	PAFZZ	PAFZZ	PAFZZ	PAFZZ	2930-01-275-1639 SHIELD, SOUND .... ABC 134-4597 (44940)	EA 1	1	
13	12	PAFZZ	PAFZZ	PAFZZ	PAFZZ	2510-01-275-5349 INSULATION BLANKET .... ABC 140-2039 (44940)	EA	1 12	
13	13	PAFZZ	PAFZZ	PAFZZ	PAFZZ	5306-01-275-5005 BOLT, SELF-LOCKING .... ABC 815-4671 (44940)	EA 8	8	
13	14	PAFZZ	PAFZZ	PAFZZ	PAFZZ	2930-01-275-4249 HOUSING, INSULATION .... ABC 134-4562 (44940)	EA 1	1	
13	15	XBHZZ	XBHZZ	XBHZZ	XBHZZ	BACKPLATE ASSEM, SC .... ABC 134-4575-1 (44940)	EA 1	1	
13	16	PAHZZ	PAHZZ	PAHZZ	PAHZZ	5305-01-276-1627 SCREW, TAPPING .... ABC 815-0627 (44940)	EA 2	2	
13	17	PAHZZ	PAHZZ	PAHZZ	PAHZZ	535-01-282-1863 POINTER, DIAL .... ABC 160-1340 (44940)	EA	1	1
13	18	PAOZZ	PAOZZ	PAOZZ	PAOZZ	5340-01-275-3526 BUTTON, PLUG .... ABC 517-0230 (44940)	A	1	

(1) ILLUSTRATION (a) FIG NO.	(b)	(2) SMR CODE a ARMY	b AIR FORCE	d NAVY	e USMC	(3) NATIONAL STOCK NUMBER	(4) DESCRIPTION USABLE ON CODE REF NUMBER MFR CODE	(5) U/M	(6) QTY INC IN	(7) USMC QTY PER EQUIP
							GROUP 05 ENGINE			
13	19	PAFZZ	PAFZZ	PAFZZ	PAFZZ	5306-01-275-3241	BOLT, MACHINE.................................. ABC 718-1027 (44940)	EA	5	6
13	20	PAFZZ	PAFZZ	PAFZZ	PAFZZ	5340-01-275-3401	BRACKET, ANGLE................................ ABC 403-2441 (449)	EA	1	2
13	21	PAFZZ	PAFZZ	PAFZZ	PAFZZ	5310-01-276-1660	WASHER, FLAT .... ABC 740-1006 (44940) C0740100600 (15434)	EA	1	12
13	22	PAFZZ	PAFZZ	PAFZZ	PAFZZ	5306-01-275-001	BOLT, MACHINE.................................. ABC 718-1037 (44940) C0718103700 (15434)	EA	1	1

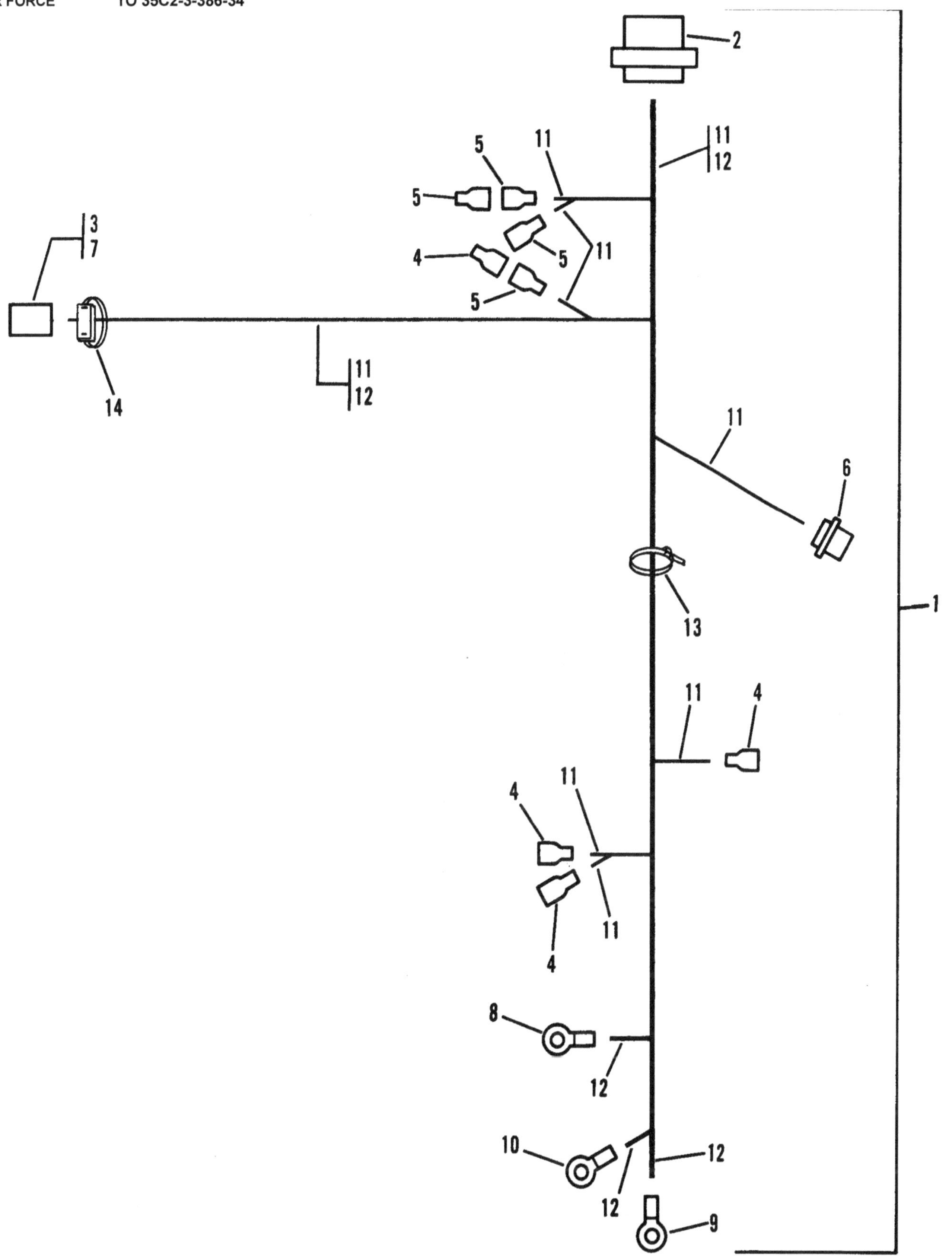

Figure 14. Engine Wiring Harness, Group 01 (DC Electrical and Control System)

(1) ILLUSTRATION (a) FIG NO.	(1) ILLUSTRATION (b) ITEM NO.	(2) SMR CODE a ARMY	(2) SMR CODE b AIR FORCE	(2) SMR CODE d NAVY	(2) SMR CODE e USMC	(3) NATIONAL STOCK NUMBER	(4) DESCRIPTION REF NUMBER MFR CODE	USABLE ON CODE	(5) U/M	(6) QTY INC IN UNITS	(7) USMC QTY PER EQUIP
							GROUP 01 DC ELECRICAL AND CONTROL SYSTEM				
14	11	MFFFF	MFFFF	MFFFF	MFFFF		HARNESS, WIRING 338-1196 (44940) 84-13118 (30554)	ABC	EA 1	1	
14	2	PAFZZ	PAFZZ	PAFZ	PAFZZ	5935-01-070-3681	CONNECTOR, PLUG MS3456W32-7S (96906)	ABC	EA 1	1	
14	3	PAFZZ	PAFZZ	PAFZZ	PAFZZ	5935-01-175-0255	CONNECTOR BODY, PLUG 6294493 (77060) 84-13120 (30554)	ABC	EA 1	1	
14	4	PAFZZ	PAFZZ	PAFZZ	PAFZZ	5935-00-115-2307	CONNECTOR, PLUG MS27144-2 {96906) 72-5017 (30554)	ABC	EA 5	5	
14	5	PAFZZ	PAFZZ	PAFZZ	PAFZZ		CONNECTOR, PLUG MS27142-3 (96906)	ABC			
14	6	PAFZZ	PAFZZ	PAFZZ	PAFZZ	5935-00-115-2306	CONNECTOR PLUG MS3456WI4SSS (96906) 323-0682 (44940)	ABC	EA 5	3	
14	7	PAFZZ	PAFZZ	PAFZZ	PAFZZ	593-00-564-5362	TERMINAL, QUICK 2965142 (77060)	ABC	EA 1	1	
14	8	PAFZZ	PAFZZ	PAFZZ	PAFZZ	5940-00-788-1586	TERMINAL LUG MS25036-158 (96906)	ABC			
14	9	PAFZZ	PAFZZ	PAFZZ	PAFZZ	5940-00-682-2445	TERMNAL, LUG MS25036-113 (96906)	ABC	EA 2	2	
14	10	PAFZZ	PAFZ	PAFZZ	PAFZZ		TERMINAL LUG MS25036-112 (96906)	ABC	EA 1	1	
14	11	PAFZZ	PAFZ	PAFZZ	PAFZZ	5940-0113-8183	WIRE, ELECTRICAL M5086/3-16-9 (81349)	ABC	EA 1	1	
14	12	PAFZZ	PAFZZ	PAFZZ	PAFZZ	594040-143-4794	WIRE, ELECTRICAL M5086/2-12-9 (81349)	ABC	EA 7	1	
14	13	PAFZZ	PAFZZ	PAFZZ	PAFZZ	6145-00-55-2562	STRAP, TIEDOWN MS3367-4-9 (96906)	ABC			
14	14	PAFZZ	PAFZZ	PAFZZ	PAFZZ	6145-00-578-7514	STRAP, TIEDOWN MS3368-1-9A (96906)	ABC	FT	16	48
						5975-00727-5153			FT	16	48
						5975-00-944-1499			EA	32	34
									A 1	1	

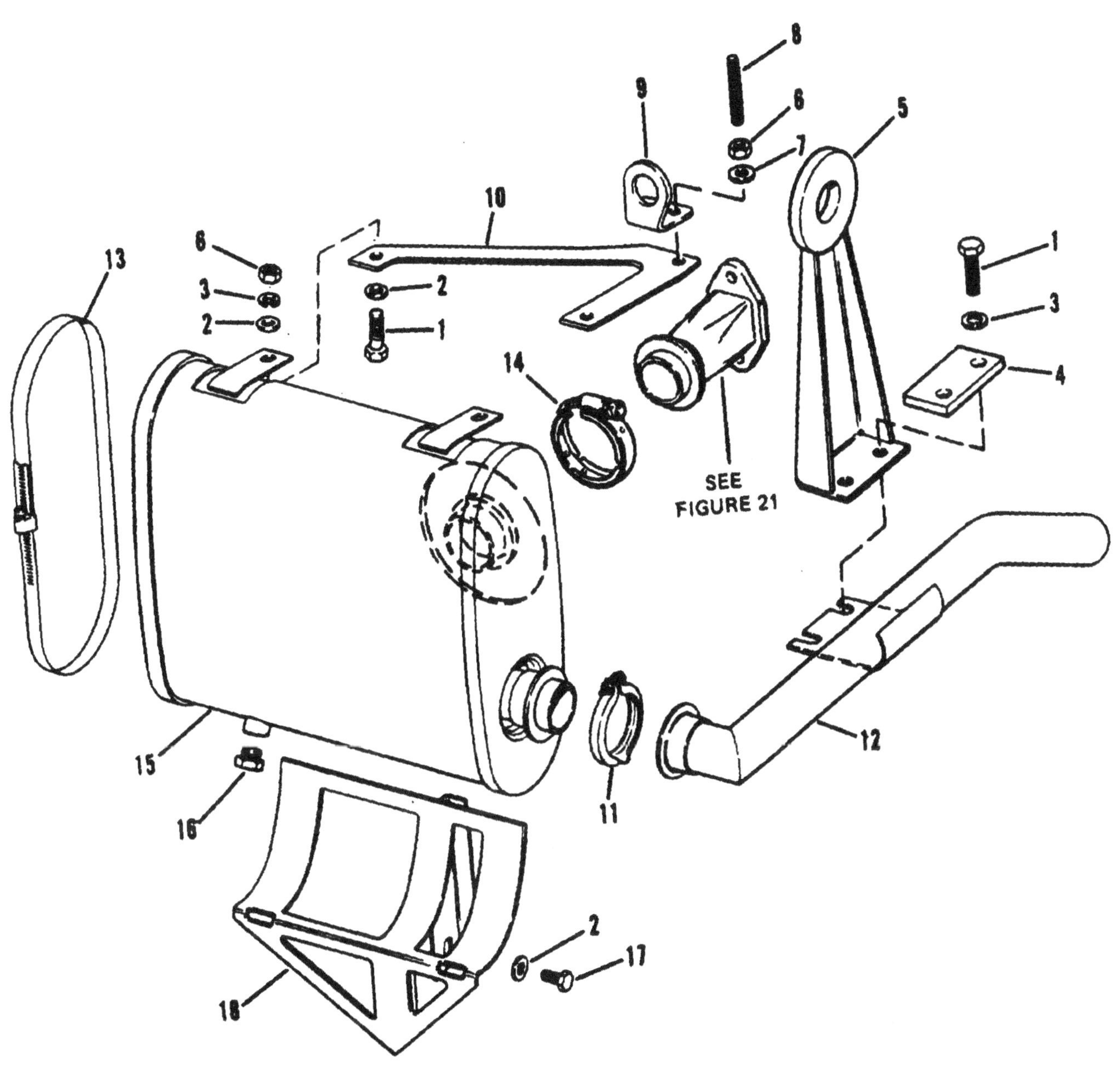

Figure 15. Engine Exhaust Assembly, Group 06 (Engine Exhaust}

(1) ILLUSTRATION (a) FIG NO.	(b)	(2) SMR CODE a ARMY	b AIR FORCE	d NAVY	e USMC	(3) (4) NATIONAL STOCK NUMBER DESCRIPTION CODE REF NUMBER MFR CODE USABLE ON	(5) U/M	(6) QTY INC IN	(7) USMC QTY PER EQUI
						GROUP 06 ENGINE EXHAUST			
15	1	PAOZZ	PAOZZ	PAOZZ	PAOZZ	5306-01-275-5002 BOLT, MACHINE.................................. ABC 800-2059 (44940)	EA 4	4	
15	2	PAOZZ	PAOZZ		PAOZZ	5310-01-275-7757 WASHER FLAT .... ABC 526-210 (44940)	EA 6	6	
15	3	PAOZZ	PAOZZ		PAOZZ	5310-00-407-9566 WASHERLOCK .... ABC MS35338-45 (96906) 850-1045 (44940)	EA 4	4	
15	4	XBOZZ	XBOZZ		XBOZZ	PLATE, SPACER.................................. ABC 403-2373 (44940) 84-13092 (30554)			
15	5	XBOZZ	XBOZZ		XBOZZ	LIFTING EYE... ABC 403-2365 (44940) 84-13030 (30554)	EA 1	1	
15	6	PAFZZ	PAFZZ		PAFZZ	5310-01-212-3338 NUT, PLAIN, HEXAGON .... ABC C0750100600 (15434)	EA 1	1	
15	7	PAFZZ	PAFZZ		PAFZZ	5310-01-276-1660 WASHER, FLAT .... ABC 740-1006 (44940) C0740100600 (15434)	EA 3	3	
15	8	PAFZZ	PAFZZ		PAFZZ	5307-0-275-3424 STUD, SHOULDERED .... ABC 520-2206 (44940)	EA 1	1	
15	9	PAFZZ	PAFZZ		PAFZZ	5340-01-275-3401 BRACKET, ANGLE............................... ABC 403-2441 (44940)			
15	10	XBFZZ	XBFZZ		XBFZZ	BRACKET, MUFFLER UP .... ABC 155-2163 (44940)	EA 3	1	
15	11	PAOZZ	PAOZZ		PAOZZ	4730-0570-2932 CLAMP, PIPE .... ABC 10906258 (19207)	EA 1	1	
15	12	PAOZZ	PAOZZ		PAOZZ	2990-01-275-1717 PIPE, EXHAUST.................................. ABC 155-2091 (44940) 84-1306 (30554)			
15	13	PAOZZ	PAOZZ		PAOZZ	4730-01-274-6700 CLAMP, HOSE .... ABC 503-1401 (44940)	EA 1	1	
15	14	PAOZZ	PAOZZ		PAOZZ	4730-01-287-8895 CLAMP, HOSE .... ABC 155-217 (44940)	EA 1	1	
15	15	PAOZZ	PAOZZ		PAOZZ	2990-01-275-1757 MUFFLER, EXHAUST .... ABC 155-2098 (44940)	A 11	1	
15 1	16	PAOZZ	PAOZZ		PAOZZ	4730-00-011-2578 .... PLUG, PIPE 502-0153 (44940) MS14314-2X (96906)	EA 2	2	
							EA 1	1	
							EA 1	1	

(1) ILLUSTRATION (a) FIG NO.	(1) ILLUSTRATION (b)	(2) SMR CODE a ARMY	(2) SMR CODE b AIR FORCE	(2) SMR CODE d NAVY	(2) SMR CODE e USMC	(3) NATIONAL STOCK NUMBER	(4) DESCRIPTION REF NUMBER MFR CODE	(4) USABLE ON CODE	(5) U/M	(6) QTY INC IN	(7) USMC QTY PER EQUIP
							GROUP 06 ENGINE EXHAUST				
15	17	PAOZZ	PAOZZ	PAOZZ	PAOZZ	5305-01-212-3220	SCREW C0800205700 (15434)	ABC	EA	2	2
15	18	XBOZZ	XBOZZ	XBOZZ	XBOZZ		SUPPORT MUFFLER............................ 155-2088 (44940)	ABC	EA	1	1

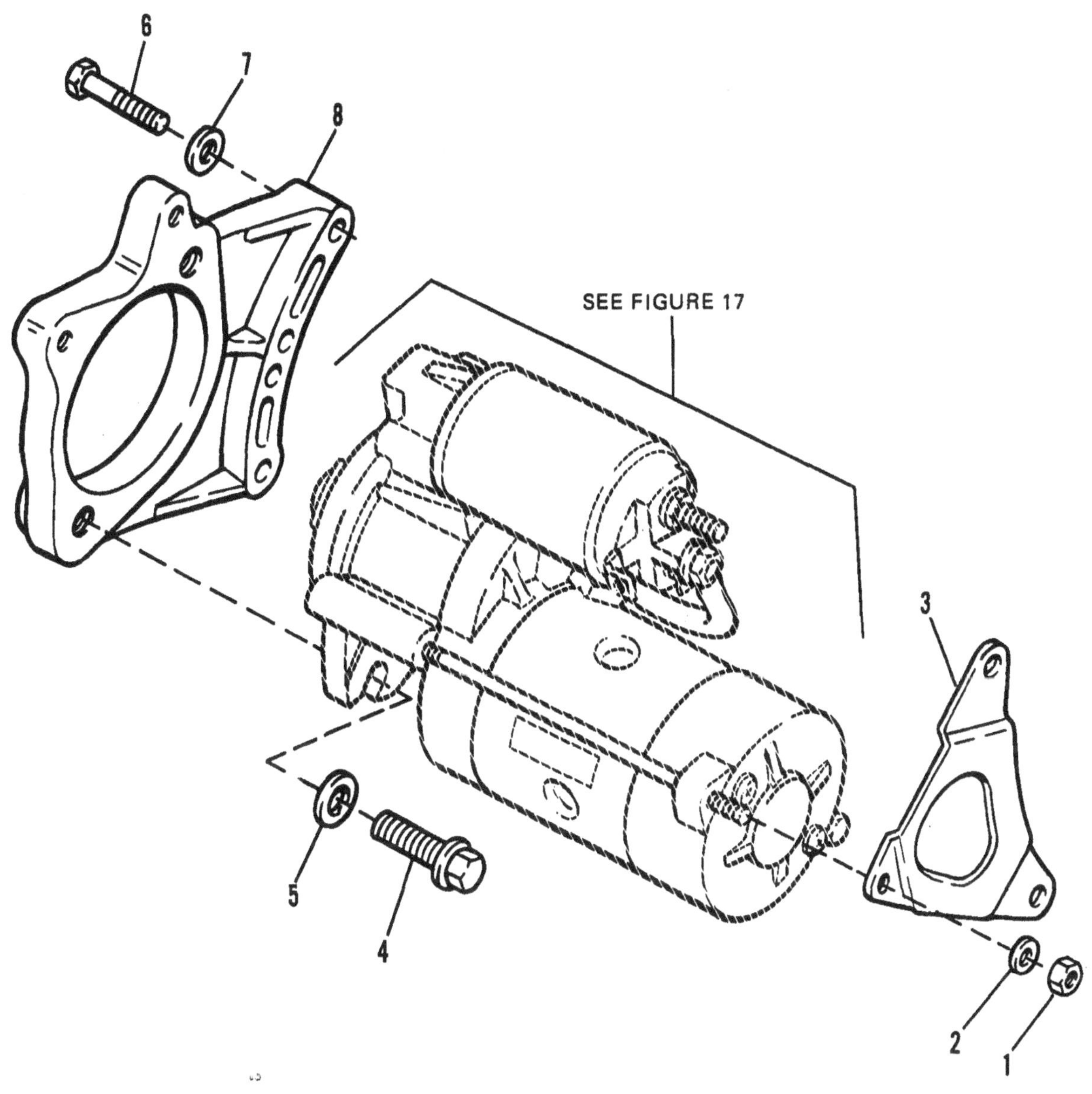

Figure 16. Starter Mounting, Group 05 (Engine)

(1) ILLUSTRATION (a) FIG NO.	(b)	(2) SMR CODE a ARMY	b AIR FORCE	d NAVY	e USMC	(3) (4) NATIONAL STOCK NUMBER DESCRIPTION REF NUMBER MFR CODE USABLE ON CODE	(5) U/M	(6) QTY INC IN	(7) USMC QTY PER EQUIP
						GROUP 05 ENGINE			
16	1	PAOZZ	PAOZZ	PAOZZ	PAOZZ	5310-01-275-3298 NUT, PLAIN, HEXAGON .... ABC 750-1004 (44940)	EA 1	1	
16	2	PAOZZ	PAOZZ	PAOZZ	PAOZZ	5310-01-275-3318 WASHER, FLAT .... ABC 740-1004 (44940)	EA 1	1	
16	3	XBOZZ	XBOZZ	XBOZZ	XBOZZ	BRACKET, STARTER .... ABC 191-1647 (44940)	EA 1	1	
16	4	PAOZZ	PAOZZ	PAOZZ	PAOZZ	5305-01-212-3227 SCREW .... ABC 720-106 (44940)	EA 2	2	
16	5	PAOZZ	PAOZZ	PAOZZ	PAOZZ	5310-0-276-8608 WASHER, FLAT .... ABC 740-1010 (44940)	EA 2	2	
16	6	PAHZZ	PAHZZ	PAHZZ	PAHZZ	5306-01-275-6001 BOLT, MACHINE.................................. ABC 718-1040 (44940)	EA 3	3	
16	7	PAHZZ	PAHZZ	PAHZZ	PAHZZ	5310-01-276-1660 WASHER, FLAT .... ABC 740-1006 (44940)	EA 3	3	
16	8	XBHZZ	XBHZZ	XBHZZ	XBHZZ	MOUNT, STARTER.............................. ABC 191-1610 (44940)	EA 1	1	

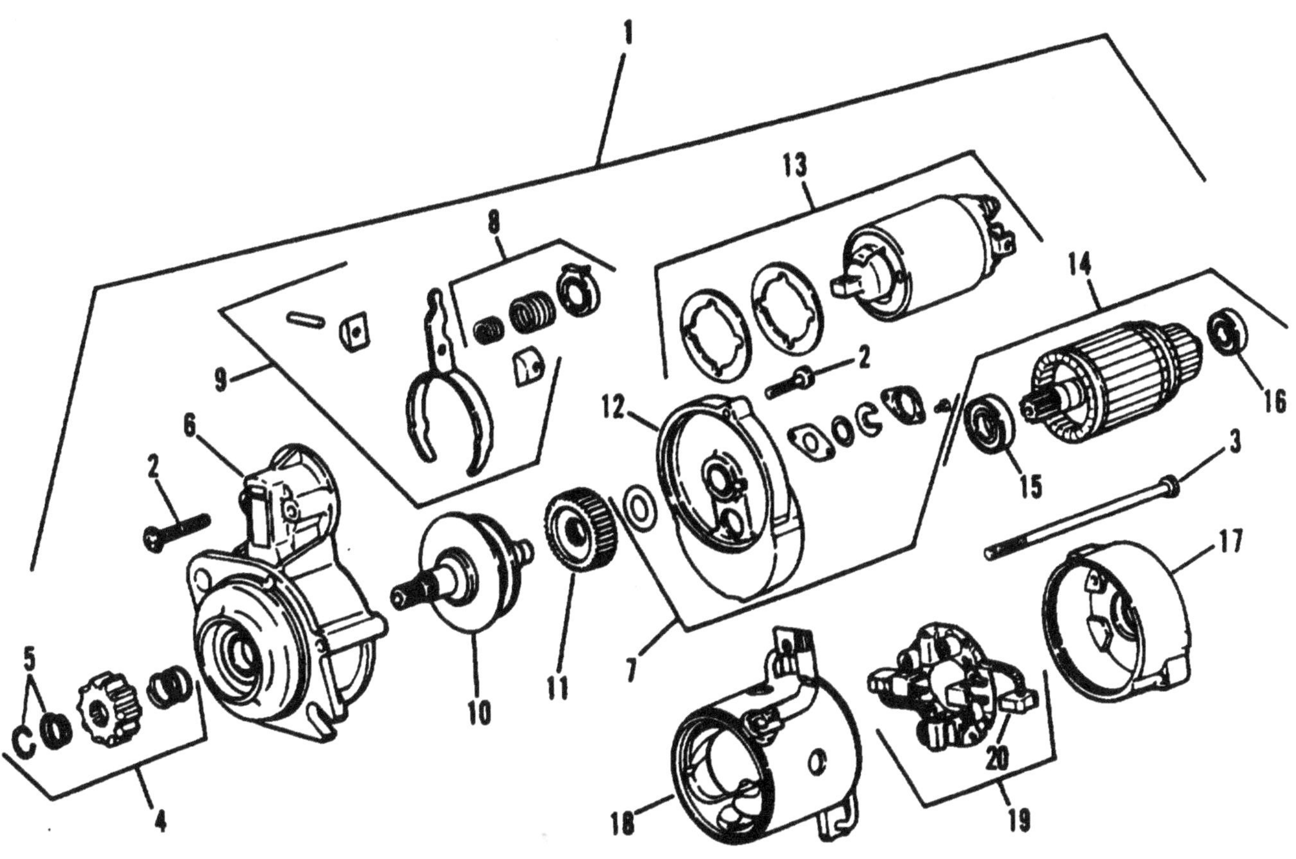

Figure 17. Starter Motor, Group 05 (Engine)

(1) ILLUSTRATION (a) FIG NO.	(b)	(2) SMR CODE a ARMY	b AIR FORCE	d NAVY	e USMC	(3) NATIONAL STOCK NUMBER	(4) DESCRIPTION REF NUMBER MFR CODE	USABLE ON CODE	(5) U/M	(6) QTY INC IN	(7) USMC QTY PER EQUIP
							GROUP 05 ENGINE				
17	1	PAOFF	PAOFF	PAOFF	PAOFF	2120-01-275-1621	STARTER, ENGINE 191-1611 (44940)	ABC	EA 1	1	
17	2	PAFZZ	PAFZZ	PAFZZ	PAFZZ	5305-01-212-3371	SCREW, SET 191-1450 (44940)	ABC	EA 3	3	
17	3	PAFZZ	PAFZZ	PAFZZ	PAFZZ	5306-01-277-3178	BOLT, MACHINE.................................. 191-1848 (44940)	ABC	EA 2	2	
17	4	PAFZZ	PAFZZ	PAFZZ	PAFZZ	2920-01-274-6728	PARTS KIT, STARTER......................... 191-1S60 (44940)	ABC	EA 1	1	
17	5	PAFZZ	PAFZZ	PAFZZ	PAFZZ	2920-01-224-4247	PARTS KIT 191-1442 (44940)	ABC	EA 1	1	
17	6	PAFZZ	PAFZZ	PAFZZ	PAFZZ	2920-01-223-8762	HOUSING... 191-1605 (44940)	ABC	EA 1	1	
17	7	PAFZZ	PAFZZ	PAFZZ	PAFZZ	2920-01-224-6246	PARTS KIT 191-1444 (44940)	ABC	EA 1	1	
17	8	PAFZZ	PAFZZ	PAFZZ	PAFZZ	5360-01-227-6302	SPRING ASSORTMENT 191-1440 (44940)	ABC	EA 1	1	
17	9	PAFZZ	PAFZZ	PAFZZ	PAFZZ	2520-01-223-8780	SHIFTER FORK 191-1441 (44940)	ABC	EA 1	1	
17	10	PAFZZ	PAFZZ	PAFZZ	PAFZZ	2920-01-274-6763	DRIVE, STARTER 191-1859 (44940)	ABC	EA 1	1	
17	11	PAFZZ	PAFZZ	PAFZZ	PAFZZ	3020-01-225-6988	GEAR 191-1446 (44940)	ABC	EA 1	1	
17	12	PAFZZ	PAFZZ	PAFZZ	PAFZZ	5340-01-223-8726	BRACKET... 191-1439 (44940)	ABC	EA 1	1	
17	13	PAFZZ	PAFZZ	PAFZZ	PAFZZ	594541-213-9233	SWITCH ASSY 191-1609 (44940)	ABC	EA 1	1	
17	14	PAFZZ	PAFZZ	PAFZZ	PAFZZ	2815-01-212-4008	ARMATURE ASSY 191-1606 (44940)	ABC	EA 1	1	
17	15	PAFZZ	PAFZZ	PAFZZ	PAFZZ	3110-01-214-0623	BEARING.CENTER............................... 191-1437 (44940)	ABC	EA 1	1	
17	16	PAFZZ	PAFZZ	PAFZZ	PAFZZ	3110-01-236-6408	BEARING 191-1436 (44940)	ABC	EA 1	1	
17	17	PAFZZ	PAFZZ	PAFZZ	PAFZZ		BRACKET, REAR................................. 191-1432 (44940)	ABC			
17	18	PAFZZ	PAFZZ	PAFZZ	PAFZZ		STATOR ASSY..................................... 191-1861 (44940)	ABC			

(1) ILLUSTRATION (a) FIG NO.	(1) ILLUSTRATION (b)	(2) SMR CODE a ARMY	(2) SMR CODE b AIR FORCE	(2) SMR CODE d NAVY	(2) SMR CODE e USMC	(3) NATIONAL STOCK NUMBER	(4) DESCRIPTION USABLE ON CODE REF NUMBER MFR CODE	(5) U/M	(6) QTY INC IN	(7) USMC QTY PER EQUIP
							GROUP 05 ENGINE			
17	19	PAFZZ	PAFZZ	PAFZZ	PAFZZ	5977-01-224-2917	HOLDER, ELECTRICAL........................ ABC 191-1601 (44940)	EA	1	1
17	20	PAFZZ	PAFZZ	PAFZZ	PAFZZ	5360-01-227-3195	SPRING, BRUSH.................................. ABC 191-1433 (44940)	EA	4	4

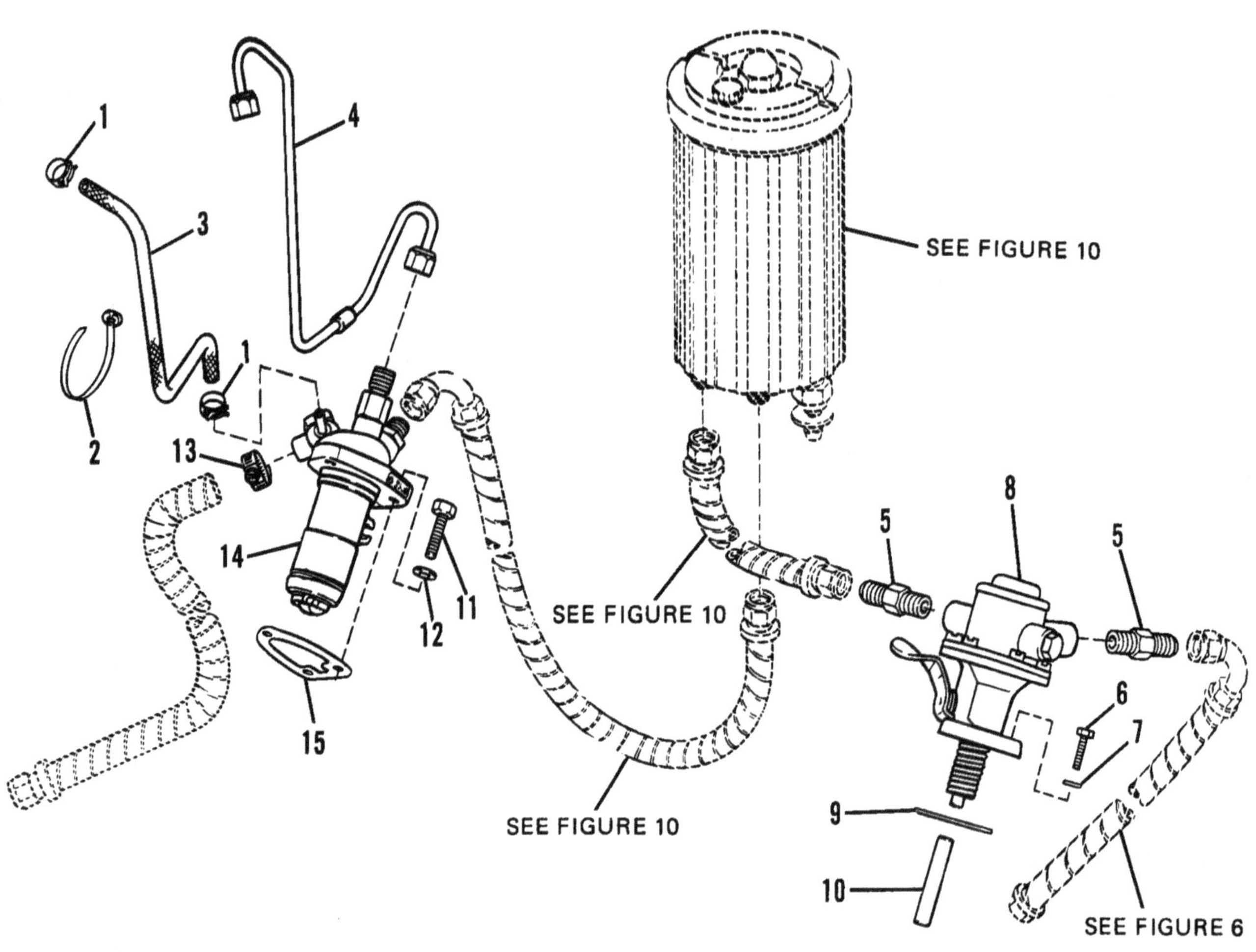

Figure 18. Fuel Transfer and Injection Pump, Group 04 (Fuel System and Oil Drain)

(1) ILLUSTRATION (a) FIG NO.	(1) ILLUSTRATION (b)	(2) SMR CODE a ARMY	(2) SMR CODE b AIR FORCE	(2) SMR CODE d NAVY	(2) SMR CODE e USMC	(3) (4) NATIONAL STOCK NUMBER / DESCRIPTION REF NUMBER MFR CODE / USABLE ON CODE	(5) U/M	(6) QTY INC IN	(7) USMC QTY PER EQUIP
						GROUP 04 FUEL SYSTEM AND OIL DRAIN			
18	1	PAFZZ	PAFZZ	PAFZZ	PAFZ	4730-01-110-9055 CLAMP, HOSE .... ABC 503-1445401 (44940)	EA	2	2
18	2	PAFZZ	PAFZZ	PAFZZ	PAFZZ	5975-01-727-5153 STRAP, TIEDON .... ABC MS336749 (9690)	EA	2	2
18	3	PAFZZ	PAFZZ	PAFZZ	PAFZZ	4720-01-212-2604 HOSE .... ABC 503-1055 (44940)	FT	AR	0
18	4	PAFZZ	PAFZZ	PAFZZ	PAFZZ	4710-01-275-531 TUBE ASSY, METAL....... ABC 1474721 (44940)	EA	1	1
18	5	PAOZZ	PAOZZ	PAOZZ	PAOZZ	4730-00-289-5484 ADAPTER, STRAIGHT....... ABC 502-101 (44940)	EA	2	2
18	6	PAOZZ	PAOZZ	PAOZZ	PAOZZ	5305-01-212-3221 SCREW .... ABC 718-1021 (44940)	EA	3	5
18	7	PAOZZ	PAOZZ	PAOZZ	PAOZZ	5310-01-275-3318 WASHER, FLAT .... ABC 740-100 (44940)	EA	3	3
18	8	PAOOO	PAOOO	PAOOO	PAOOO	2910-01-276-1483 PUMP, FUEL TRANSFER .... ABC 149-2105 (44940)	EA	1	1
18		PAFZZ	PAFZZ	PAFZZ	PAFZZ	2910-01-275-9144 ; PARTS KIT, FUEL .... ABC 149-2142 (44940) PARTS OF KUT ARE NSS	EA	1	2
18	9	PAOZZ	PAOZZ	PAOZZ	PAOZZ	5330-01-276-0897 GASKET....... ABC 149-2060 (44940) PART OF KT 160184 (44940)	EA	1	2
18	10	PAO7Z	PAOZZ	PAOZZ	PAOZZ	3040-01-275-2537 CONNECTING LINK....... ABC 149-1864 (44940)	EA	1	1
18	11	PAFZZ	PAFZZ	PAFZZ	PAFZZ	5306-01-275-6002 BOLT, MACHINE....... ABC 720-1037 (44940)	EA	3	3
18	12	PAFZZ	PAFZZ	PAFZZ	PAFZZ	5310-01-276-1660 WASHER, FLAT .... ABC 740-1006 (44940)	EA	3	3
18	13	PAFZZ	PAFZZ	PAFZZ	PAFZZ	4730-00-871-6729 CLAMP .... ABC 6202 (81646)	EA	1	1
18	14	PAFZZ	PAFZZ	PAFZZ	PAFZZ	2910-01-274-6767 PUMP, FUEL METERING....... ABC 1470714 (44940)	EA	1	1
18	15	PAFZZ	PAFZZ	PAFZZ	PAFZZ	5365-01-275-7814 SHIM....... ABC 147-40541 (44940)	EA	1	1
18	15	PAFZ	PAFZZ	PAFZZ	PAFZZ	5365-01-275-7817 SHIM....... ABC 147-0405-02 (44940)	EA	1	1

(1) ILLUSTRATION (a) FIG NO.	(1) ILLUSTRATION (b)	(2) SMR CODE a ARMY	(2) SMR CODE b AIR FORCE	(2) SMR CODE d NAVY	(2) SMR CODE e USMC	(3) (4) NATIONAL STOCK NUMBER / DESCRIPTION REF NUMBER MFR CODE / USABLE ON CODE	(5) U/M	(6) QTY INC IN	(7) USMC QTY PER EQUIP
						GROUP 04 FUEL SYSTEM AND OIL DRAIN			
18	15	PAFZZ	PAFZZ	PAFZZ	PAFZZ	5365-01-275-7818 SHIM....... ABC 1470405-03 (44940)	EA	1	1
18	15	PAFZZ	PAFZZ	PAFZZ	PAFZZ	5365-01-275-7819 SHIM....... ABC 147-04544 (44940)	EA	1	1
18		PAOZZ	PAOZZ	PAOZZ	PAOZZ	2910-01-275-9144 PARTS KIT, FUEL .... ABC 149-2142 (44940) PARTS OF KIT ARE NSS	EA	1	1

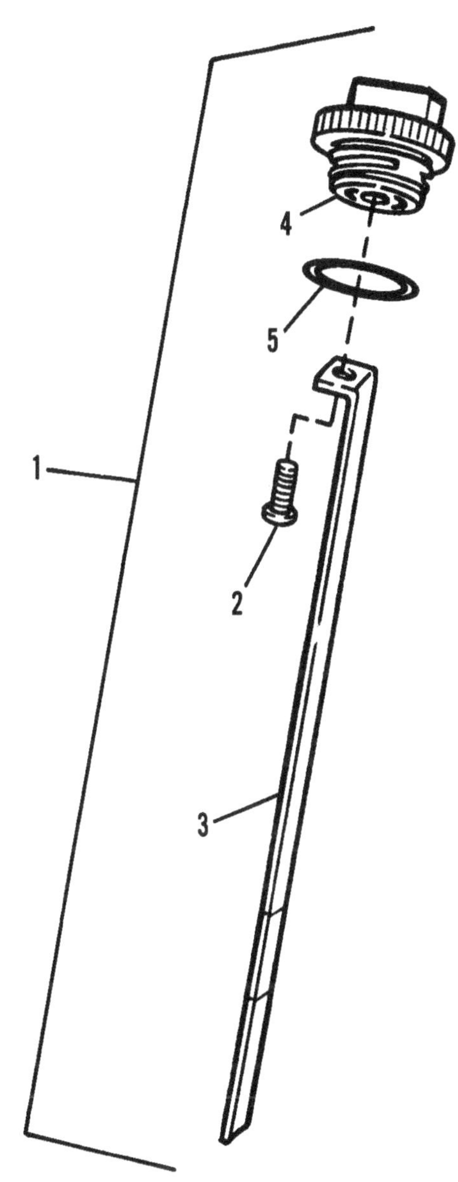

Figure 19. Oil Indicator, Group 05 (Engine)

(1) ILLUS-TRA-TION (a) FIG NO.	(1) ILLUS-TRA-TION (b)	(2) SMR CODE a ARMY	(2) SMR CODE b AIR FORCE	(2) SMR CODE d NAVY	(2) SMR CODE e USMC	(3) NATIONAL STOCK NUMBER	(4) DESCRIPTION REF NUMBER MFR CODE	(4) USABLE ON CODE	(5) U/M	(6) QTY INC IN	(7) USMC QTY PER EQUIP
							GROUP 05 ENGINE				
19	1	PAO7Z	PAOZZ	PAOZZ	PAOZZ	6680-01-271-2683	GAGE, ROD-CAP.................................. 123-169001 (44940)	ABC	EA	1	1
19	2	XAOZZ	XAOZZ	XAOZZ	XAOZZ		SCREW 115-0333 (44940)	ABC	EA	1	1
19	3	XAOZZ	XAOZZ	XAOZZ	XAOZZ		OIL LEVEL INDICATOR 123-170601 (44940)	ABC	EA	1	1
19	4	PAOZZ	PAOZZ	PAOZZ	PAOZZ	5340-01-329-3933	CAP, PROTECTIVE............................... 123-1647 (44940)	ABC	EA	1	1
19	5	PAO2Z	PAOZZ	PAOZZ	PAOZZ	5330-01-229-8077	PACKING, PREFORMED...................... 509-0142 (44940)	ABC	EA	1	1

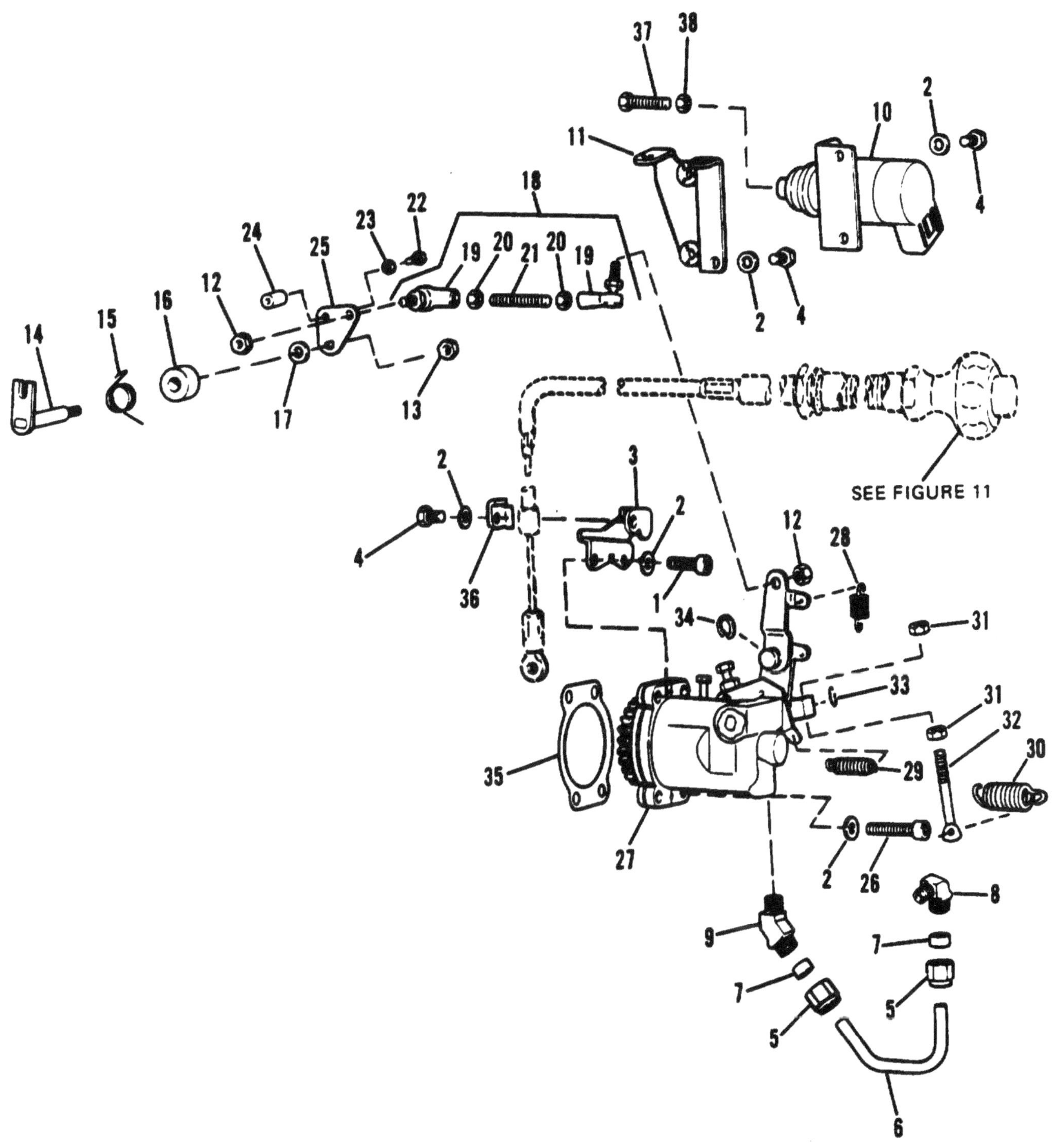

Figure 20. Governor and Fuel Shutdown Solenoid, Group 05 (Engine)

(1) ILLUSTRATION (a) FIG NO.	(b)	(2) SMR CODE a ARMY	b AIR FORCE	d NAVY	e USMC	(3) (4) NATIONAL STOCK NUMBER DESCRIPTION CODE REF NUMBER MFR CODE USABLE ON	(5) U/M	(6) QTY INC IN	(7) USMC QTY PER EQUIP
						GROUP 05 ENGINE			
20	1	PAOZZ	PAOZZ	PAOZZ	PAOZZ	5305-01-277-4990 SCREW, CAP, SOCKET .... ABC 725-1029 (44940)	A 2	2	
20	2	PAOZZ	PAOZZ	PAOZZ	PAOZZ	5310-01-275-3310 WASHER, FLAT .... ABC 740-1004 (44940)	EA 8	8	
20	3	XBOZZ	XBOZZ	XBOZZ	XBOZZ	BRACKET, CABLE MTG .... ABC 152-42 (44940)			
20	4	PAOZZ	PAOZZ	PAOZZ	PAOZZ	5306-01-275-6000 BOLT, MACHINE.................................. ABC 7111018 (44940)	EA 1	1	
20	5	PAOZZ	PAOZZ	PAOZZ	PAOZZ	5310-01-275-3301 NUT, PLAIN, CLINCH............................ ABC 50249S2 (44940)	EA 4	4	
20	6	PAOZZ	PAOZZ	PAOZZ	PAOZZ	4710-01-274-8341 TUBE, BENT, METALLIC .... ABC 120-1142 (44940)	EA 2	2	
20	7	PAOZZ	PAOZZ	PAOZZ	PAOZZ	5365-01-282-2825 BUSHING, NONMETALLIC .... ABC 502-0953 (44940) PART OF KIT 168-014 (44940)	EA 1	1	
20	8	PAOZZ	PAOZZ	PAOZZ	PAOZZ	4730-01-275-4175 ELBOW .... ABC 502-0951 (44940) VS179-VL-4-2B (93061)	EA 4	2	
20	9	PAOZZ	PAOZZ	PAOZZ	PAOZZ	473-01-275-4176 ELBOW .... ABC 502-0977 (44940) 169VL-4-2 (93061)			
20	10	PAOZZ	PAOZZ	PAOZZ	PAOZZ	5945-01-280-5477 SOLENOID .... ABC 307-2386 (44940)	EA 1	1	
20	1	XBOZZ	XBOZZ	XBOZZ	XBOZZ	BRACKET, SOLENOID .... ABC 307-2387 (44940)	EA 1	1	
20	12	PAOZZ	PAOZZ	PAOZZ	PAOZZ	5310-01-049-2745 NUT, PLAIN, ASSEMBLED .... ABC 10-0131 (44940)			
20	13	PAOZZ	PAOZZ	PAOZZ	PAOZZ	5310-01-276-1640 NUT, SELF-LOCKING .... ABC 870-2052 (44940)	EA 1	1	
20	14	PAOZZ	PAOZZ	PAOZZ	PAOZZ	3040-01-275-1640 SHAFT, SHOULDERED .... ABC 152127 (44940)	EA 1	1	
20	15	PAOZZ	PAOZZ	PAOZZ	PAOZZ	5360-01-275-3516 SPRING, HELICAL .... ABC 150-2132 (44940)	EA 3	3	
20	16	PAOZZ	PAOZZ	PAOZZ	PAOZZ	5365-01-275-6873 SPACER SLEEVE .... ABC 150-2135 (44940)			
20	17	PAOZZ	PAOZZ	PAOZZ	PAOZ	5310-01-275-3319 FLAT WASHER .... ABC 150-2130 (44940)	EA 1	1	
20	18	PAOOO	PAOOO	PAOOO	PAOOO	3040-01-275-0203 BALL JOINT, DOUBLE .... ABC 1502044 (44940)	EA 1	1	
							EA 1	1	
							EA 1	1	

(1) ILLUSTRATION (a) FIG NO.	(1) ILLUSTRATION (b)	(2) SMR CODE a ARMY	(2) SMR CODE b AIR FORCE	(2) SMR CODE d NAVY	(2) SMR CODE e USMC	(3) NATIONAL STOCK NUMBER	(4) DESCRIPTION CODE REF NUMBER MFR CODE	USABLE ON	(5) U/M	(6) QTY INC IN	(7) USMC QTY PER EQUIP
							GROUP 05 ENGINE				
20	19	PAOZZ	PAOZZ	PAOZZ	PAOZZ	3040-01-049-0578	JOINT, BALL 1500939 (440)	ABC	EA	2	2
20	20	PAOZZ	PAOZZ	PAOZZ	PAOZZ	5310-00-934-9751	NUT, PLAIN, HEXAGON MS35650-302 (96906)	ABC	EA	2	2
20	21	PAOZZ	PAOZZ	PAOZZ	PAOZZ	3040-01-275-0205	CONNECTING LINK 150-2045 (4494)	ABC	EA	1	1
20	22	PAOZZ	PAOZZ	PAOZZ	PAOZZ	5305-01-277-3184	SCREW, MACHINE 8004074 (44940)	ABC	EA	1	1
20	23	PAOZZ	PAOZZ	PAOZZ	PAOZZ	5310-01-275-3327	LOCKWASHER 850-2004 (44940)	ABC	EA	1	1
20	24	PAOZZ	PAOZZ	PAO	PAOZ	5310-01-277-7335	NUT, PLAIN, ROUND 150-204 (44940)	ABC	EA	1	1
20	25	XBOZZ	XBOZZ	XBOZZ	XBOZZ		EXTERNAL LINK 150-2126 (44940)	ABC	EA	1	1
20	26	PAOZZ	PAOZZ	PAOZZ	PAOZZ	5305-01-276-1637	SCREW, CAP, SOCKET 725-1031 (44940)	ABC	EA	2	2
20	27	PAOZZ	PAOZZ	PAOZZ	PAOZZ	2910-01-275-1715	GOVERNOR, DIESEL 150-2129 (44940) GD1033G (74465)	ABC	EA	1	1
20	28	PAFZZ	PAFZZ	PAFZZ	PAFZZ	5360-01-275-3513	SPRING, HELICAL 150-2192 (44940)	ABC	EA	1	1
20	29	PAFZZ	PAFZZ	PAFZZ	PAFZZ	5360-01-275-3514	SPRING, HELICAL 150-222S (44940)	ABC	EA	1	1
20	30	PAFZZ	PAFZZ	PAFZZ	PAFZZ	5360-01-277-1192	SPRING, HELICAL 102229 (44940)	ABC	EA	1	1
20	31	PAFZZ	PAFZZ	PAFZZ	PAFZZ	5310-00-043-0520	NUT, PLAIN, HEXAGON MS356sO3252 (906)	ABC	EA	2	2
20	32	PAFZZ	PAFZZ	PA'ZZ	PAFZZ	5306-01-278-1963	BOLT EYE 150-2230 (44940)	ABC	EA	1	1
20	33	PAFZZ	PAFZZ	PAFZZ	PAFZZ	5365-01-274-9748	RING, RETAINING 518-0034 (44940)	ABC	EA	1	1
20	34	PAFZZ	PAFZZ	PAFZZ	PAFZZ	5365-01-173-3442	RING, RETAINING 518-0207 (44940)	ABC	EA	1	1
		PAOZZ	PAOZZ	PAOZZ	PAOZZ	5330-01-276-7501	GASKET 10-2133 (44940) PART OF KT1680184 (44940)	ABC	EA	1	2
		PAOZZ	PAOZZ	PAOZZ	PAOZZ	5340-01-275-3528	STRAP, RETAINING 152-0260 (44940)	ABC	EA	1	1

(1) ILLUSTRATION (a) FIG NO.	(1) (b)	(2) SMR CODE a ARMY	b AIR FORCE	d NAVY	e USMC	(3) NATIONAL STOCK NUMBER	(4) DESCRIPTION USABLE ON CODE REF NUMBER MFR CODE	(5) U/M	(6) QTY INC IN	(7) USMC QTY PER EQUIP
							GROUP 05 ENGINE			
20	37	PAOZZ	PAOZZ	PAOZZ	PAOZZ	5305-01-275-3291	SCREW, CAP, HEXAGON .... ABC 815-0633 (44940)	A	1	1
20	36	PAOZZ	PAOZZ	PAOZZ	PAOZZ	5310-00-869-1018	NUT, PLAIN, HEXAGON .... ABC 871-0018 (44940) MS35650-3255 (96906)	EA	1	1

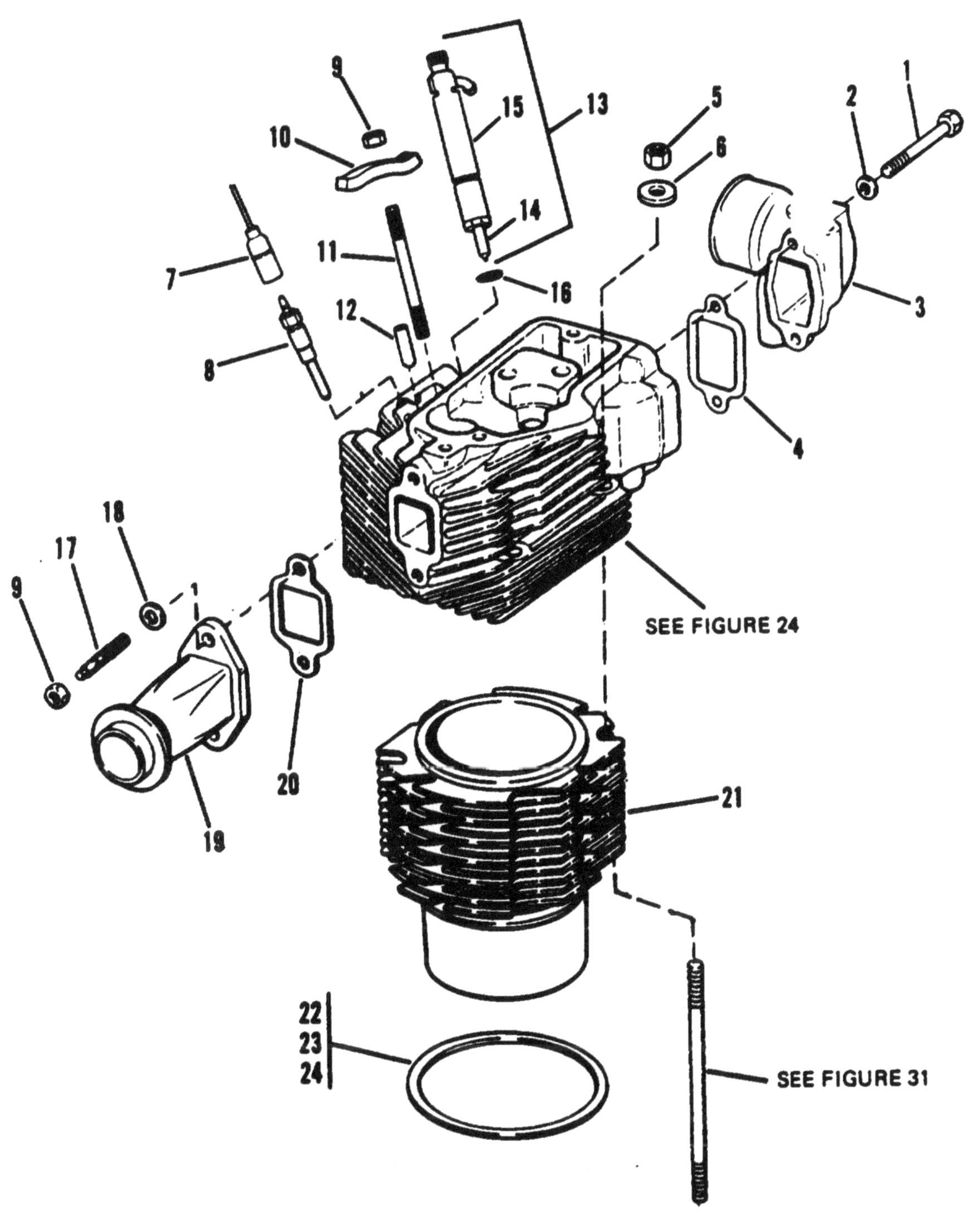

Figure 21. Cylinder, Group 05 (Engine)

(1) ILLUSTRATION (a) FIG NO.	(1) ILLUSTRATION (b)	(2) SMR CODE a ARMY	(2) SMR CODE b AIR FORCE	(2) SMR CODE d NAVY	(2) SMR CODE e USMC	(3) NATIONAL STOCK NUMBER	(4) DESCRIPTION REF NUMBER MFR CODE	(4) USABLE ON CODE	(5) U/M	(6) QTY INC IN	(7) USMC QTY PER EQUIP
							GROUP 05 ENGINE				
21	1	PAOZZ	PAOZZ	PAOZZ	PAOZZ	5306-01-275-3242	BOLT, MACHINE 718-1048 (44940) C0718104800 (15434)	ABC	EA 2	2	
21	2	PAOZZ	PAOZZ	PAOZZ	PAOZZ	5310-01-276-1660	WASHER, FLAT 7401006 (44940)	ABC	EA 2	2	
21	3	PAOZZ	PAOZZ	PAOZZ	PAOZZ	2815-01-274-9474	MANIFOLD INTAKE 154-2636 (44940)	ABC	EA 1	1	
21	4	PAOZZ	PAOZZ	PAOZZ	PAOZZ	5330-01-275-3358	GASKET 154-23U (44940) PART OF KIT 168-0184 (44940)	ABC	EA 2	1	
21	5	PAFZZ	PAFZZ	PAFZZ	PAFZZ	5310-01-275-3299	NUT, PLAIN, HEXAGON 110-3219 (44940)	ABC	EA 4	4	
21	6	PAFZZ	PAFZZ	PAFZZ	PAFZZ	5310-01-275-3320	WASHER, FLAT 52640313 (44940)	ABC	EA 4	4	
21	7	PAOZZ	PAOZZ	PAOZZ	PAOZZ	6150-01-274-5935	LEAD, ELECTRICAL 226-3310 (44940)	ABC	EA 1	1	
21	8	PAOZZ	PAOZZ	PAOZZ	PAOZZ	2920-01-275-4311	PLUG, GLOW 333-0242 (44940) 19G (70040)	ABC	EA 1	1	
21	9	PAFZZ	PAFZZ	PAFZZ	PAFZZ	5310-01-137-4829	NUT, PLAIN, HEXAGON 860-2058 (44940) 430-008 (72741)	ABC	EA 3	3	
21	10	PAFZZ	PAFZZ	PAFZZ	PAFZZ	5340-01-276-5912	CLAMP, RIM CLENCHING 147-0555 (44940)	ABC	EA 1	1	
21	11	PAFZZ	PAFZZ	PAFZZ	PAFZZ	5307-01-275-3425	STUD 520-22O (44940)	ABC	EA 1	1	
21	12	PAFZZ	PAFZZ	PAFZZ	PAFZZ	5315-01-276-9216	PIN, STRAIGHT 7754076 (44940)	ABC	EA 1	1	
21	13	PAFZZ	PAFZZ	PAFZZ	PAFZZ	2910-01-274-9375	NOZZLE, FUEL 147-0715-01 (44940) 0432 191 797 (53867)	ABC	EA 1	1	
21	14	PAFZZ	PAFZZ	PAFZZ	PAFZZ		TIP, NOZZLE 1474094 (44940) 0433 117 120 (53867)	ABC	EA 1	1	
21	15	PAFZZ	PAFZZ	PAFZZ	PAFZZ		WASHER ASSORTMENT 147-0793 (44940)	ABC	EA 1	1	
21	15	XBFZZ	XBFZZ	XBFZZ	XBFZZ		SHIM- 1.00 2-430-101-035 (53867) 147-0793-1 (44940)	ABC	EA 1	1	
21	15	XBFZZ	XBFZZ	XBFZZ	XBFZZ		SHIM- 1.04 2-430101-037 (53867) 147-0793-02 (44940)	ABC	EA 1	1	

(1) ILLUSTRATION (a) FIG NO.	(b)	(2) SMR CODE a ARMY	b AIR FORCE	d NAVY	e USMC	(3) NATIONAL STOCK NUMBER	(4) DESCRIPTION USABLE ON CODE REF NUMBER MFR CODE	(5) U/M	(6) QTY INC IN	(7) USMC QTY PER EQUIP
							GROUP 05 ENGINE			
21 15		XBFZZ	XBFZZ	XBFZZ	XBFZZ		SHIM- 1.08 .... ABC 2-430-101-039 (53867) 1474793403 (44940)	EA	1	1
21 15		XBFZZ	XBFZZ	XBFZZ	XBFZZ		SHIM-1.12 .... ABC 2-430-1014-041 (53867) 14747934-04 (44940)	EA	1	1
21 15		XBFZZ	XBFZZ	XBFZZ	XBFZ		SHIM- 1.16 .... ABC 2-430-101443 (53867) 14747934- (44940)	EA	1	1
21 15		XBFZZ	XBFZZ	XBFZZ	XBFZZ		SHIM- 1.20 .... ABC 2-430-101045 (53867) 1474793-06 (44940)	EA	1	1
21 15		XBFZZ	XBFZZ	XBFZZ	XBFZZ		SHIM- 1.24 .... ABC 2-430-101447 (53867) 1474793-047 (44940)	EA	1	1
21 15		XBFZZ	XBFZZ	XBFZZ	XBFZ		SHIM- 1.28 .... ABC 2-430-101-449 (53867) 1474793-8 (44940)	EA	1	1
21 15		XBFZZ	XBFZZ	XBFZZ	XBFZZ		SHIM- 1.32 .... ABC 2-430-101451 (53867) 1474793-09 (44940)	EA	1	1
21 15		XBFZZ	XBFZZ	XBFZZ	XBFZZ		SHIM- 1.36 .... ABC 2-430-1014O3 (53867) 1474793-10 (44940)	EA	1	1
21 15		XBFZZ	XBFZZ	XBFZZ	XBFZZ		SHIM- 1.40 .... ABC 2-43101-55 (53867) 1470793-11 (44940)	EA	1	1
21 15		XBFZZ	XBFZZ	XBFZZ	XBFZZ		SHIM- 1.44 .... ABC 2-430-101417 (53867) 1474793-12 (44940)	EA	1	1
21 15		XBFZZ	XBFZZ	XBFZZ	XBFZZ		SHIM- 1.48 .... ABC 2-430101-49 (53867) 147-793-13 (44940)	EA	1	1
21 15		XBFZZ	XBZZ	XBFZZ	XBFZZ		SHIM- 1.52 .... ABC 2430-101-061 (53867) 1474793-14 (44940)	EA	1	1
21 15		XBFZZ	XBFZZ	XBFZZ	XBFZZ		SHIM- 1.56 .... ABC 2-430-101063 (53867) 1470793-15 (44940)	EA	1	1
21 15										
21 15										

(1) ILLUSTRATION (a) FIG NO.	(b)	(2) SMR CODE a ARMY	b AIR FORCE	d NAVY	e USMC	(3) (4) NATIONAL STOCK NUMBER / DESCRIPTION / CODE / REF NUMBER MFR CODE / USABLE ON	(5) U/M	(6) QTY INC IN	(7) USMC QTY PER EQUIP
						GROUP 05 ENGINE			
21	15	XBFZZ	XBFZZ	XBFZZ	XBFZZ	SHIM- 1.60 .... ABC 2-430-101065 (53867) 1474793-16 (44940)	EA 1	1	
21	15	XBFZZ	XBFZZ	XBFZZ	XBFZZ	SHIM- 1.64 .... ABC 2-430-101-067 (53867) 14740793-17 (44940)	EA 1	1	
21	15	XBFZZ	XBFZZ	XBFZZ	XBFZZ	SHIM- 1.68 .... ABC 2-430-101-69 (53567) 1474793-18 (44940)	EA 1	1	
21	15	XBFZZ	XBFZZ	XBFZZ	XBFZZ	SHIM- 1.72 .... ABC 2-430-101-071 (53867) 1474793-19 (44940)	EA 1	1	
21	15	XBFZZ	XBFZZ	XBFZ	XBFZZ	SHIM- 1.76 .... ABC 2-430-1014-73 (53867) 1474793-20 (44940)	EA 1	1	
21	15	XBFZZ	XBFZZ	XBFZZ	XBFZZ	SHIM- 1.80 .... ABC 2-430-101475 (53867) 1474793-21 (44940)	EA 1	1	
21	15	XBFZZ	XBFZZ	XBFZZ	XBFZZ	SHIM- 1.84 .... ABC 2-430-101-077 (53867) 147-0793-22 (44940)	EA 1	1	
21	15	XBZZ	XBFZZ	XBFZZ	XBFZZ	SHIM- 1.88 .... ABC 2-430-101-079 (53867) 1474793-23 (44940)	EA 1	1	
21	15	XBFZZ	XBFZZ	XBFZZ	XBFZZ	SHIM- 1.92 .... ABC 2-430-1014-1 (53867) 1474793-24 (44940)	EA 1	1	
21	15	XBFZZ	XBFZZ	XBFZZ	XBFZZ	SHIM- 1.96 .... ABC 2-430-101-083 (53867) 147-793-25 (44940)	EA 1	1	
21	16	PAFZZ	PAFZZ	PAFZZ	PAFZZ	5310-01-275-3321 WASHER, FLAT .... ABC 147-0430 (44940) PART OF KIT 16-0184 (44940)	EA 1	1	
21	17	PAFZZ	PAFZZ	PAFZZ	PAFZZ	5307-01-275-3425 STUD, PLAIN .... ABC 520-2202 (44940)	EA 2	1	
21	18	PAFZZ	PAFZZ	PAFZZ	PAFZZ	5310-O1-276-1660 WASHER, FLAT .... ABC 740-1006 (44940)	EA 2	2	
21	19	PAFZZ	PAFZZ	PAFZZ	PAFZZ	2990-01-274-9376 PIPE, EXHAUST .... ABC 154-2646 (44940)	EA 2	2	

(1) ILLUSTRATION (a) FIG NO.	(1) ILLUSTRATION (b)	(2) SMR CODE a ARMY	(2) SMR CODE b AIR FORCE	(2) SMR CODE d NAVY	(2) SMR CODE e USMC	(3) NATIONAL STOCK NUMBER	(4) DESCRIPTION REF NUMBER MFR CODE	USABLE ON CODE	(5) U/M	(6) QTY INC IN	(7) USMC QTY PER EQUIP
							GROUP 05 ENGINE				
21	20	PAFZZ	PAFZZ	PAFZZ	PAFZZ	5330-01-275-3359	GASKET.. 154-2342 (44940) PART OF KIT 1680114 (44940)	ABC	EA	1	2
221	1	PAFHH	PAFHH	PAFHH	PAFHH	2815-01-274-9369	BARREL, CYLWDER............................ 110-3312 (44940)	ABC	EA	1	1
21	22	PAFZZ	PAFZZ	PAFZZ	PAFZZ	5365-01-215-6846	SHIM, 05.. 110-2725.01 (44940)	ABC	EA	1	1
21	23	PAFZZ	PAFZZ	PAFZZ	PAFZZ	5365-01-275-6947	SHIM .15.. 110-2725-2 (44940)	ABC	EA	1	1
21	24	PAFZZ	PAFZZ	PAFZZ	PAFZZ	5365-01-275-6848	SHIM 35... 110-2725-03 (44940)	ABC	RA	1	1

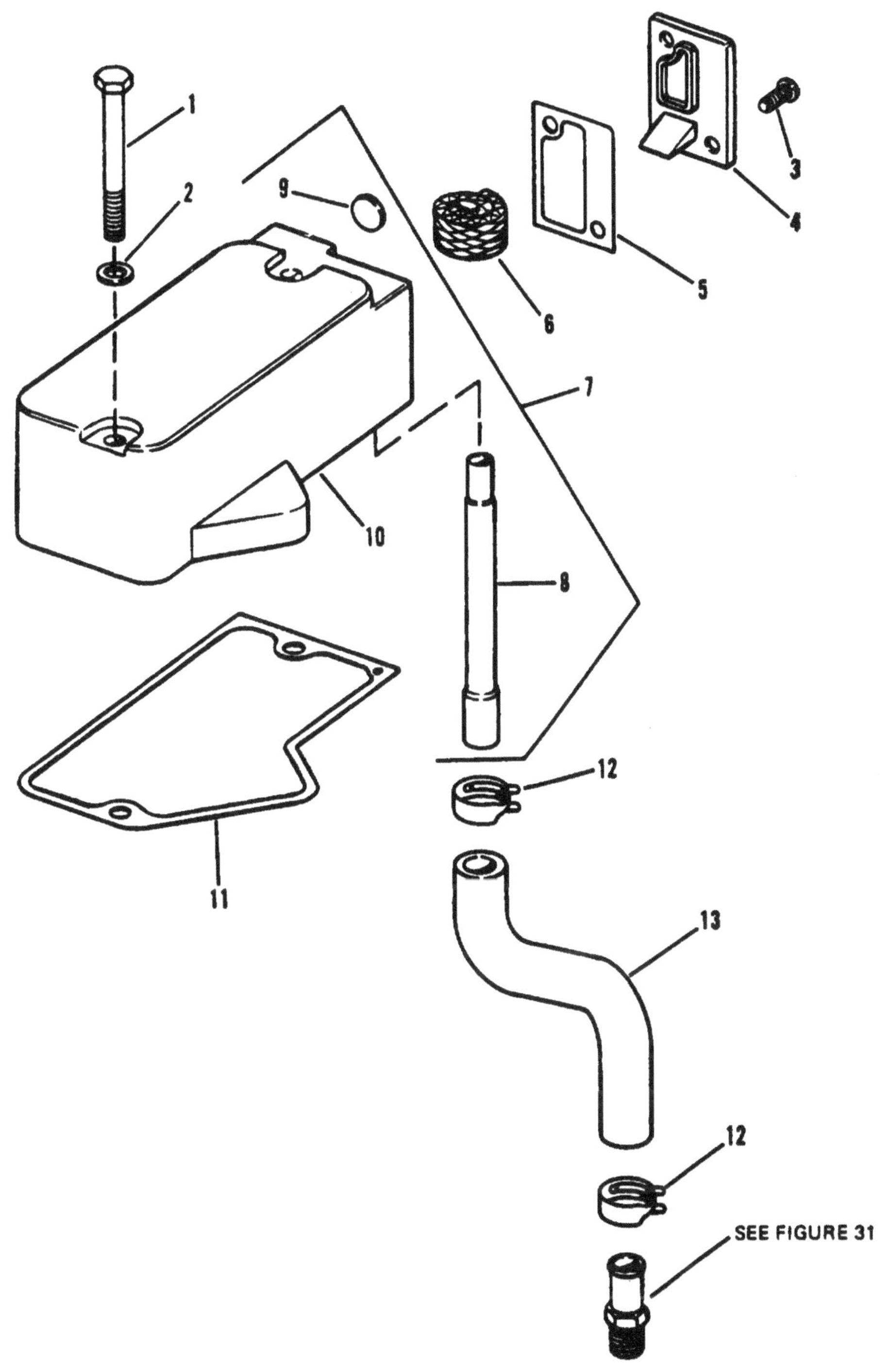

Figure 22. Rocker Arm Cover and Breather, Group 05 (Engine)

(1) ILLUSTRATION (a) FIG NO.	(b)	(2) SMR CODE a ARMY	b AIR FORCE	d NAVY	e USMC	(3) (4) NATIONAL STOCK NUMBER DESCRIPTION CODE REF NUMBER MFR CODE USABLE ON	(5) U/M	(6) QTY INC IN	(7) USMC QTY PER EQUIP
						GROUP 05 ENGINE			
22	1A	PAOZZ	PAOZZ	PAOZZ	PAOZZ	5305-01-226-6624 SCREW .... ABC 718-1046 (44940)	EA 2	2	
22	2	PAOZZ	PAOZZ	PAOZZ	PAOZZ	5310-01-282-5S92 WASHER, FLAT .... ABC 740-1 8 (44940)	EA 2	2	
22	3	PAOZZ	PAOZZ	PAOZZ	PAOZZ	5305-01-276-1628 SCREW .... ABC S815638 (44940)	EA 2	2	
22	4	PAOZZ	PAOZZ	PAOZZ	PAOZZ	534-01-305-3406 COVER, ACCESS .... ABC 123-17(0 (44940)	EA 1	1	
22	5	PAOZZ	PAOZZ	PAOZZ	PAOZZ	5330-01-275-3360 GASKET........ ABC 1154400 (44940) PART OF KIT 161-0184 (44940)	EA 2	1	
22	6	PAOZZ	PAOZZ	PAOZZ	PAOZZ	2940-01-214-8456 FILTER, ELEMENT........ ABC 123-1707 (44940) PART OF KIT 11684 (44940)	EA 2	1	
22	7	XBOZZ	XBOZZ	XBOZZ	XBOZZ	COV. ASSY ROCKER ARM .... ABC 115-0402 (44940)	EA 1	1	
22	8	XBOZZ	XBOZZ	XBOZZ	X'BOZZ	TUBE, BREATHER........ ABC 123-169S (44940)	EA 1	1	
22	9	XBOZZ	XBOZZ	XBOZZ	XBOZZ	PLUG, WELCH .... ABC 5174235 (44940)	EA 1	1	
22	10	XBOZZ	XBOZZ	XBOZZ	XBOZZ	COVER, ROCKER ARM........ ABC 115-0353 (44940)	EA 1	1	
22	11	PAOZZ	PAOZZ	PAOZZ	PAOZZ	5330-01-275-3361 GASKET........ ABC 115-0351 (4940) PART OF KIT 16.0184 (44940)	EA 2	1	
22	12	PAOZZ	PAOZZ	PAOZZ	PAOZZ	4730-00-908-3194 CLAMP, HOSE .... ABC MS35842-11 (96906)	EA 2	2	
22	13	PAOZ	PAOZZ	PAOZZ	PAOZZ	4720-01-274-4932 HOSE, PREFORMED........ ABC 503-1440 (44940)	EA 1	1	

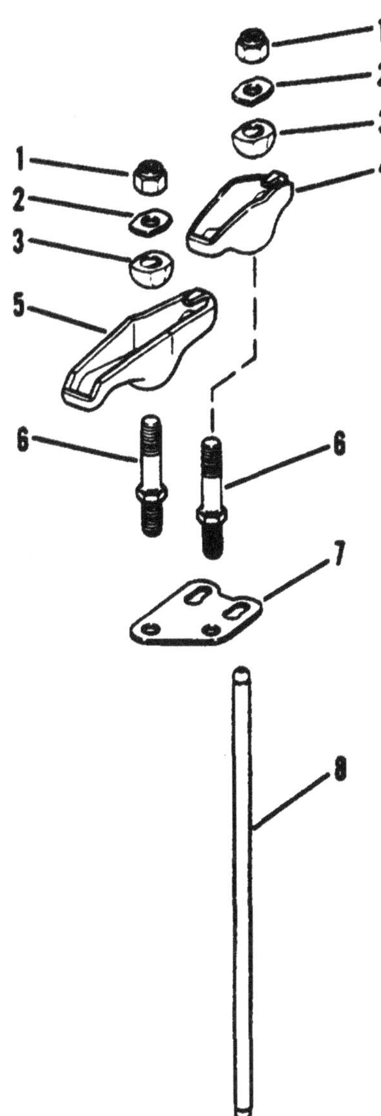

Figure 23. Rocker Arms and Push Rods, Group 05 (Engine)

(1) ILLUSTRATION (a) FIG NO.	(b)	(2) SMR CODE a ARMY	b AIR FORCE	d NAVY	e USMC	(3) NATIONAL STOCK NUMBER	(4) DESCRIPTION USABLE ON CODE REF NUMBER MFR CODE	(5) U/M	(6) QTY INC IN	(7) USMC QTY PER EQUIP
							GROUP 05 ENGINE			
23	1	PAFZZ	PAFZZ	PAFZZ	PAFZZ	531001-227-6099	NUT .... ABC 1150315 (44940)	EA 2	2	
23	2	PAFZZ	PAFZZ	PAFZZ	PAFZZ	5310-01-275-3310	WASHER, FLAT .... ABC 5262133 (44940)	EA 2	2	
23	3	PAFZZ	PAFZZ	PAFZZ	PAFZZ	2815-01-275-2484	BALL ROCKER ARM........ ABC 11504350 (44940)	EA 2	2	
23	4	PAFZZ	PAFZZ	PAFZZ	PAFZZ	2815-01-274-6768	ROCKER ARM........ ABC 115-0358 (44940)	EA 1	1	
23	5	PAFZZ	PAFZZ	PAFZZ	PAFZZ	2815-01-274-9386	ROCKER ARM........ ABC 1154359 (44940)	EA 1	1	
23	6	PAFZZ	PAFZZ	PAFZZ	PAFZZ	5307-01-277-1165	STUD, SHOULDERED .... ABC 1151211 (44940)	EA 2	2	
23	7	PAFZZ	PAFZZ	PAFZZ	PAFZZ	2815-01-275-0178	GUIDE, PUSHROD........ ABC 115-0399 (44940)	EA 1	1	
23	8	PAFZZ	PAFZZ	PAFZZ	PAFZZ	2815-01-274-9516	PUSHROD........ ABC 115-4354-01 (44940)	EA 2	2	

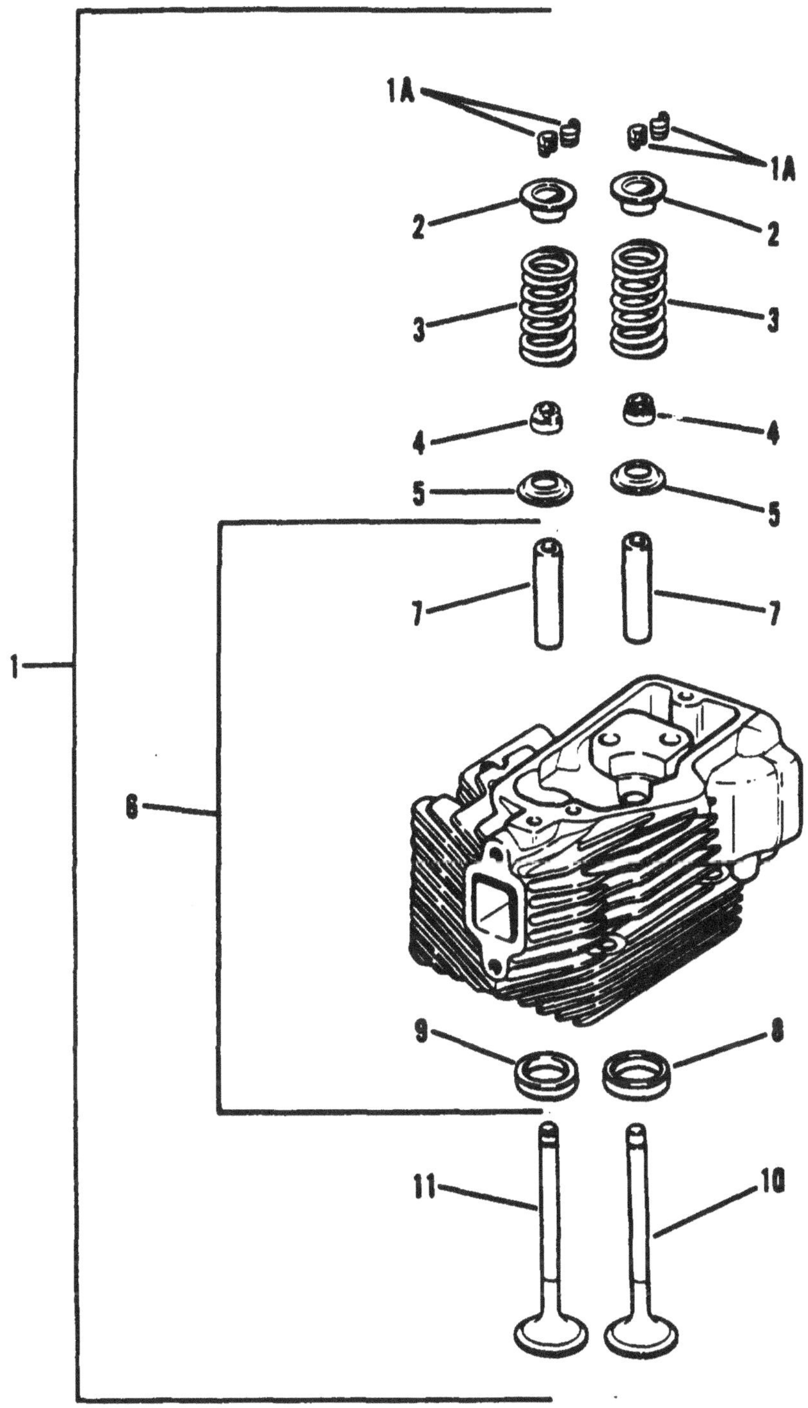

Figure 24. Cylinder Head and Valves, Group 05 (Engine)

(1) ILLUSTRATION (a) FIG NO.	(1) ILLUSTRATION (b)	(2) SMR CODE a ARMY	(2) SMR CODE b AIR FORCE	(2) SMR CODE d NAVY	(2) SMR CODE e USMC	(3) NATIONAL STOCK NUMBER	(4) DESCRIPTION USABLE ON CODE / REF NUMBER MFR CODE	(5) U/M	(6) QTY INC IN	(7) USMC QTY PER EQUIP
							GROUP 05 ENGINE			
24	1	PAFFF	PAFFF	PAFFF	PAFFF	2815-01-275-1712	CYLINDER HEAD ASSY .... ABC 110-3530 (44940)	EA	1	1
24	1A	PAFZZ	PAFZZ	PAFZZ	PAFZZ	2815-01-275-4331	LOCK, VALVE SPRING......................... ABC 110-2611 (44940)	EA	4	4
24	2	PAFZZ	PAFZZ	PAFZZ	PAFZZ	2815-01-275-5368	LOCK, VALVE SPRING......................... ABC 110-3179 (44940)	EA	2	2
24	3	PAFZZ	PAFZZ	PAFZZ	PAFZZ	5340-01-275-3506	SPRING, HELICAL .... ABC 110-3493 (44940)	EA	2	2
24	4	PAFZZ	PAFZZ	PAFZZ	PAFZZ	2590-01-274-9241	RING, WIPER.. ABC 5094221 (44940) PART OF KIT 1684-014 (44940)	EA	4	2
24	5	PAFZZ	PAFZZ	PAFZZ	PAFZZ	5340-01-276-1795	RETAINER, HELICAL............................ ABC 110-2627 (44940)	EA	2	2
24	6	PAFFF	PAFFF	PAFFF	PAFFF	2815-01-274-9368	CYLINDER HEAD.................................. ABC 110-3192 (44940)	EA	1	1
24	7	PAFZZ	PAFZZ	PAFZZ	PAFZZ	2815-01-274-6814	VALVE GUIDE....................................... ABC 110-3176 (44940)	EA	2	2
24	8	PAFZZ	PAFZZ	PAFZZ	PAFZZ	2815-01-274-8213	INSERT, ENGINE.................................. ABC 1103178 (44940)	EA	1	1
24	9	PAFZZ	PAFZZ	PAFZZ	PAFZZ	2815-01-274-8214	INSERT, ENGINE.................................. ABC 110-3177 (44940)	EA	1	1
24	10	PAFZZ	PAFZZ	PAFZZ	PAFZZ	2815-01-274-9254	VALVE, POPPET, ENGINE .... ABC 1103504 (44940)	EA	1	1
24	11	PAFZZ	PAFZZ	PAFZZ	PAFZZ		VALVE, POPPET, ENGINE .... ABC 1103505 (44940)	EA	1	1

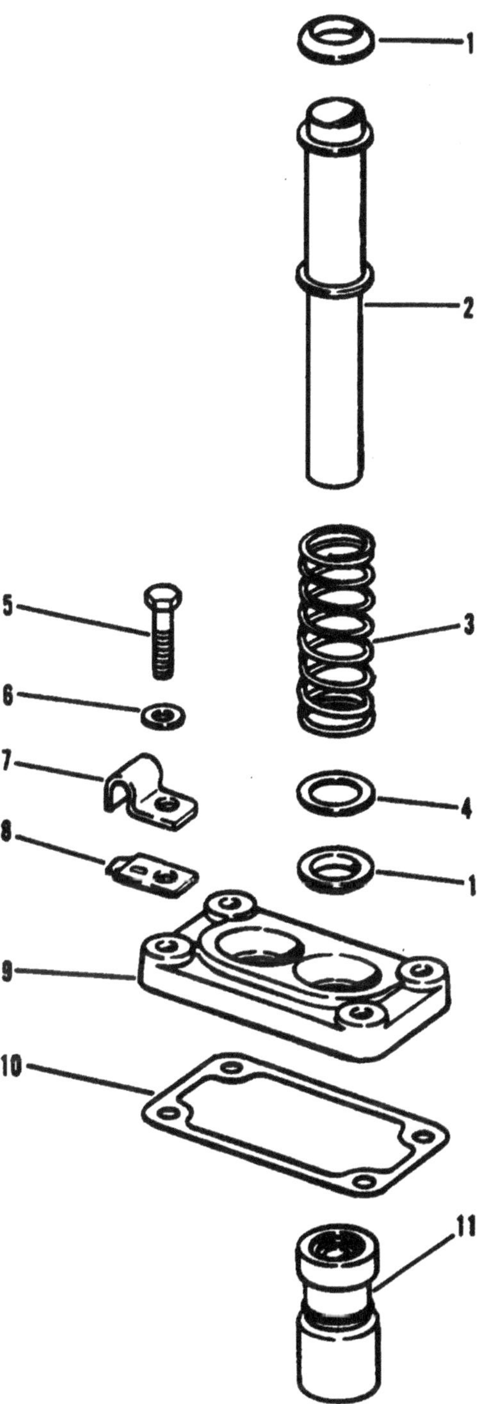

Figure 25. Push Rod Tube and Valve Tappet, Group 05 (Engine)

(1) ILLUSTRATION (a) FIG NO.	(1) ILLUSTRATION (b)	(2) SMR CODE a ARMY	(2) SMR CODE b AIR FORCE	(2) SMR CODE d NAVY	(2) SMR CODE e USMC	(3) NATIONAL STOCK NUMBER	(4) DESCRIPTION REF NUMBER MFR CODE	(4) USABLE ON CODE	(5) U/M	(6) QTY INC IN	(7) USMC QTY PER EQUIP
							GROUP 05 ENGINE				
25	1	PAFZZ	PAFZZ	PAFZZ	PAFZZ	5330-01-275-3338	PACKING, PREFORMED 1154261 (44940) PART OF KT 168-0184 (44940)	ABC	EA 4		8
25	2	XBFZZ	XBFZZ	XBFZZ	XBFZZ		TUBE, PUSH ROD 1154360 (44940)	ABC	EA 2		2
25	3	XBFZ	XBFZZ	XBFZZ	XBFZZ		SPRING PUSHROD TUBE 115-0393 (44940)	ABC	EA		2
25	4	PAFZZ	PAFZZ	PAFZZ	PAFZZ	5310-01-275-3322	WASHER, FLAT 11S293 (44940)	ABC	2		2
25	5	PAFZZ	PAFZZ	PAFZZ	PAFZZ		SCREW, CAP, HEXAGON 718-1022 (44940)	ABC	EA 2		4
25	6	PAFZZ	PAFZZ	PAFZZ	PAFZZ	5305-01-275-3287	WASHER, FLAT 740-1004 (44940)	ABC	EA 4		
25	7	PAFZZ	PAFZZ	PAFZZ	PAFZZ	5310-01-275-3318	STRAP, RETAINING 147-0759 (44940)	ABC	REA	4	1
25	8	PAFZZ	PAFZZ	PAFZZ	PAFZZ	5340-01-276-5908	CLAMP, INJECTOR LINE 147-0760 (44940)	ABC EA	EA l		
25	9	XBFZZ	XBFZZ	XBFZZ	XBFZZ	5340-01-277-1185	ADAPERPUSHRODTUB 115-I324 (44940)	ABC			1
25	10	PAFZZ	PAFZZ	PAFZZ	PAFZZ		GASKET 1154403 (44940) PART OF KRT 160184 (44940)	ABC	1 1		2
25	11	PAFZZ	PAFZZ	PAFZZ	PAFZZ		TAPPET, ENGINE 15-0397 (44940)	ABC	EA 1		2
						5330-01-276-2290			EA 1		
									EA 2		

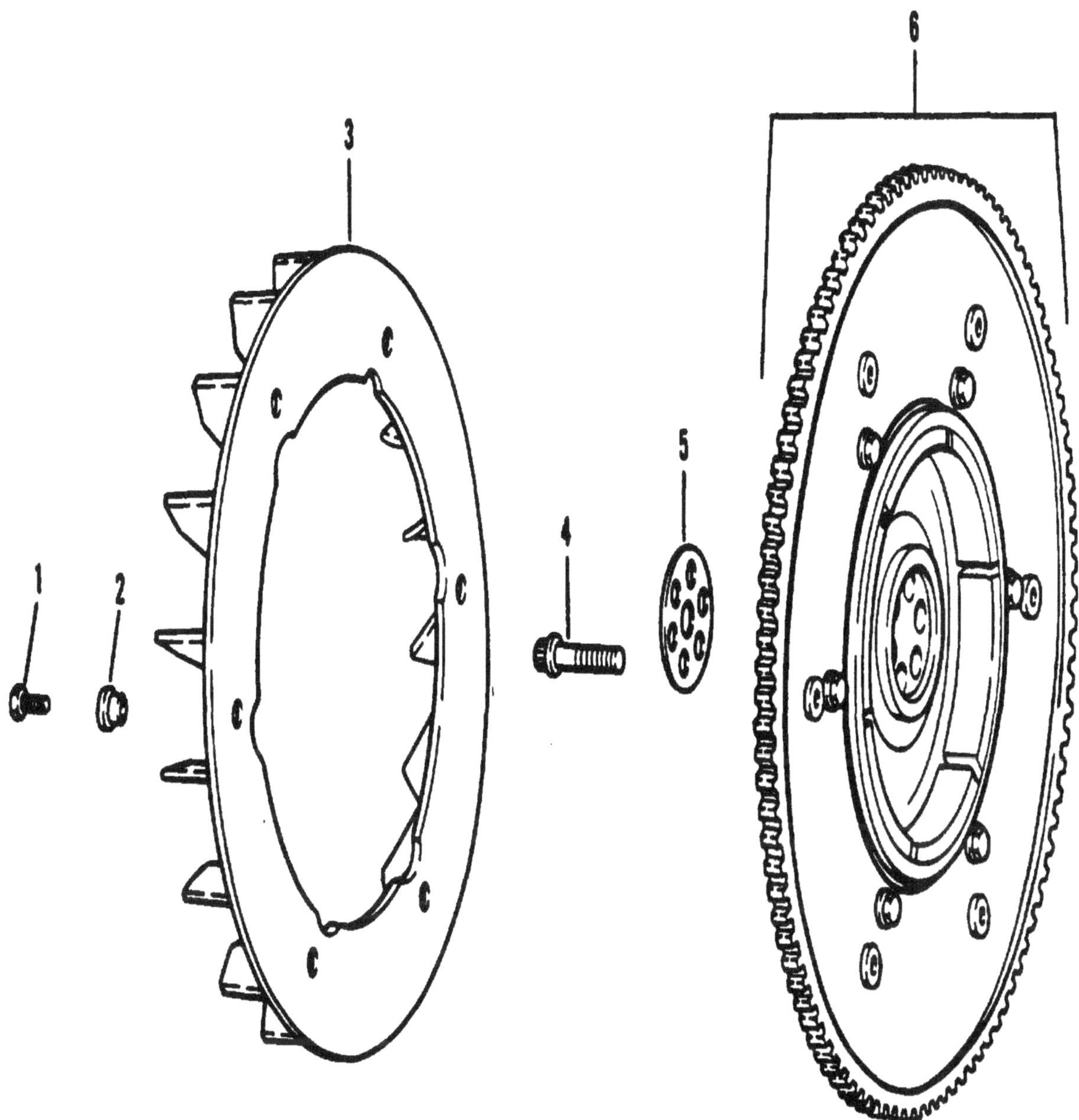

Figure 26. Flywheel and Engine Fan, Group 05 (Engine)

(1) ILLUSTRATION (a) FIG NO.	(1) ILLUSTRATION (b)	(2) SMR CODE a ARMY	(2) SMR CODE b AIR FORCE	(2) SMR CODE d NAVY	(2) SMR CODE e USMC	(3) NATIONAL STOCK NUMBER	(4) DESCRIPTION REF NUMBER MFR CODE	(4) USABLE ON CODE	(5) U/M	(6) QTY INC IN	(7) USMC QTY PER EQUIP
							GROUP 05 ENGINE				
26	1	PAFZZ	PAFZ	PAFZZ	PAFZZ	5306-01-276-7463	BOLT, MHCHINE 718-1020 (44940)	ABC	EA 6		7
26	2	PAFZZ	PAFZZ	PAFZZ	PAFZZ	3120-01-276-8639	BUSHING, SLEEVE 1344531 (44940)	ABC	EA 6		6
26	3	PAFZZ	PAFZZ	PAFZZ	PAMZ		IMPELLER, FAN 134-461 (44940)	ABC			1
26	4	PAFZZ	PAFZZ	PAFZZ	PAFZZ	4140-01-274-7721	BOLT, MACHIE 806-2004 (44940)	ABC	EA 1		6
26	5	XBFZZ	XBFZ	XBFZ	XBFZZ	530-01-275-3240	WASHERPLATE 104-1635 (44940)	ABC	EA 6		
26	6	PAFZZ	PAFZZ	PAFZZ	PAFZZ		FLYWHEEL, ENGINE 104-1643 (44940)	ABC	EA 1		1
						2815-01-274-6764			EA 1		

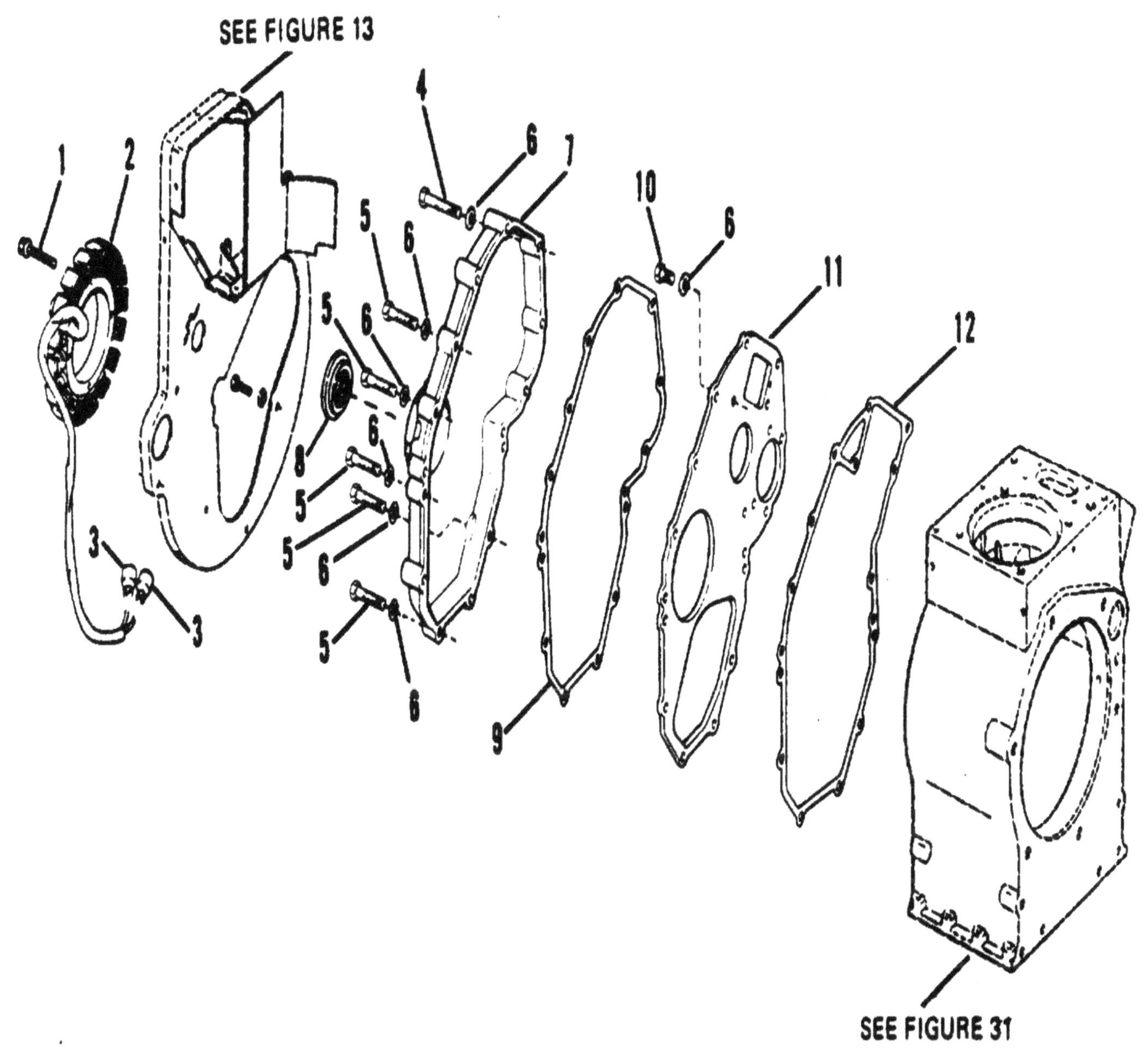

Figure 27. Crankcase Gear Cover and Stator, Group 05 (Engine)

(1) ILLUS-TRA-TION (a) FIG NO.	(b)	(2) SMR CODE a ARMY	b AIR FORCE	d NAVY	e USMC	(3) NATIONAL STOCK NUMBER	(4) DESCRIPTION REF NUMBER MFR CODE	USABLE ON CODE	(5) U/M	(6) QTY INC IN	(7) USMC QTY PER EQUIP
							GROUP 05 ENGINE				
27	1	PAFZZ	PAFZZ	PAFZZ	PAFZZ	5305-01-292-91I1	SCREW, TAPPING 815-0599 (44940)	ABC	EA 4		4
27	2	PAFZZ	PAFZZ	PAFZZ	PAFZZ	2920-01-274-9479	STATOR, ENGINE 191-1612 (44940)	ABC	EA I		I
27	3	PAFZZ	PAFZZ	PAFZZ	PAFZZ	5935-00-115-2306	CONNECTOR, PLUG MS27142-3 (96906)	ABC	EA 2		2
27	4	PAFZZ	PAFZZ	PAFZZ	PAFZZ	5306-01-275-3241	BOLT, MACHINE 71S-1027 (44940)	ABC	EA I		1
27	5	PAFZZ	PAFZZ	PAFZZ	PAFZZ	5306-01-277-3179	BOLT, MACHINE 718-1025 (44940)	ABC	EA 5		5
27	6	PAFZZ	PAFZZ	PAFZ	PAFZZ	5310-01-275-3318	WASHER FLAT 740-1004 (44940)	ABC	EA 7		7
27	7	XBFZZ	XBZZ	XBFZZ	XBFZZ	5330-01-275-5013	GEAR COVER 103-0763 (44940)	ABC	EA 1		1
27	8	PAFZZ	PAFZZ	PAFZZ	PAFZZ	5330-01-275-5014	SEAL, PLAIN ENCASED 509-0166 (44940) PART OF KIT 168-0184 (44940)	ABC	EA 1		2
27	9	PAFZZ	PAFZZ	PAFZZ	PAFZZ	5306-01-274-7463	GASKET 103-07S2 (44940) PART OF KIT 16S-0184 (44940)	ABC	EA I		2
27	10	PAFZZ	PAFZZ	PAFZZ	PAFZZ		BOLT, MACHINE 718-1020 (44940)	ABC	EA I		1
27	11	XBFZZ	XBFZZ	XBFZZ	XBFZZ		BACKPLATE 103-767 (44940)	ABC	EA I		1
27	12	PAFZZ	PAFZZ	PAFZZ	PAMZ		GASKET 1034-783 (44940) PART OF KIT 168-0184 (44940)	ABC	EA 1		2

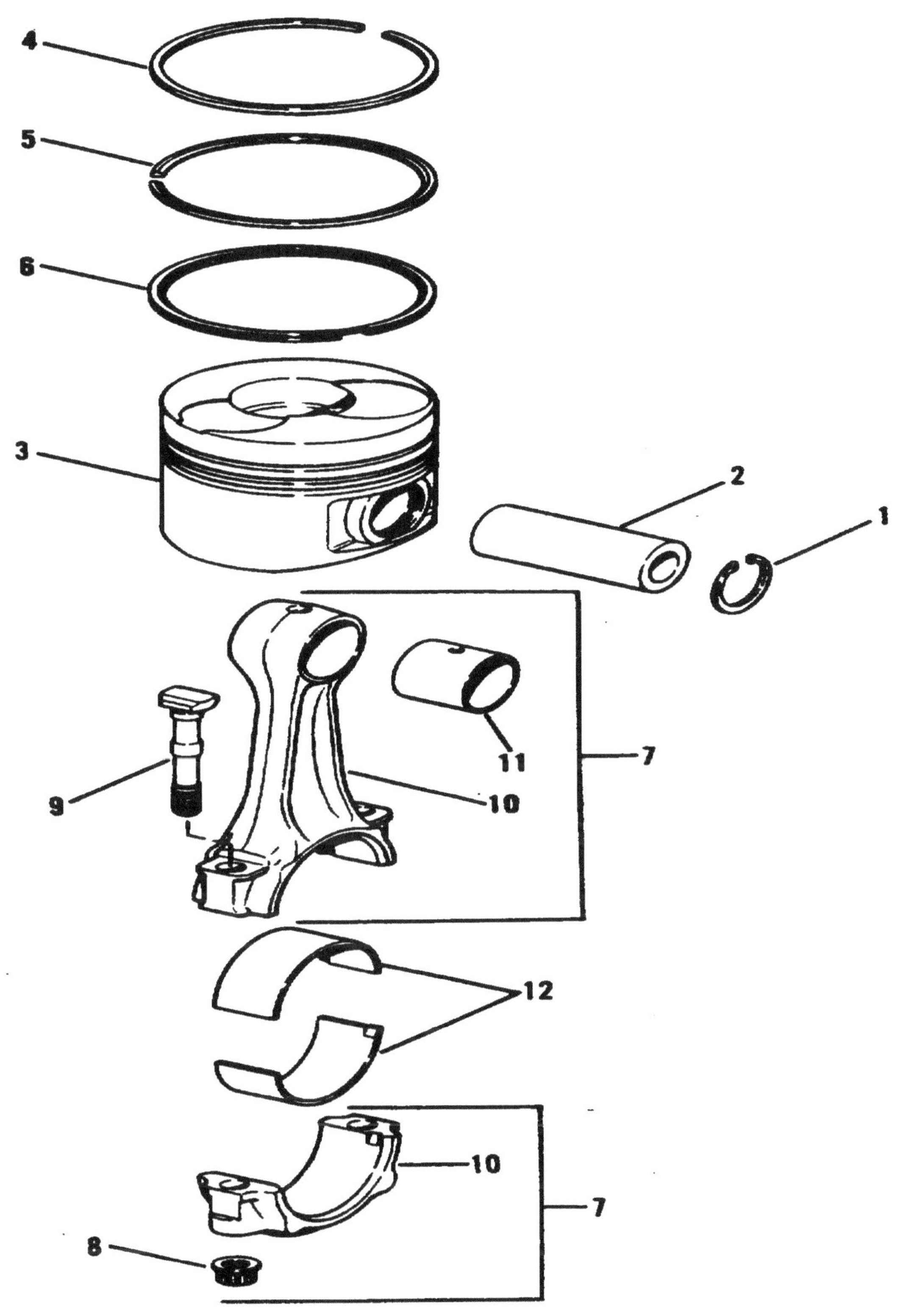

Figure 28. Connecting Rod and Piston, Group 05 (Engine)

(1) ILLUSTRATION (a) FIG NO.	(1) ILLUSTRATION (b)	(2) SMR CODE a ARMY	(2) SMR CODE b AIR FORCE	(2) SMR CODE d NAVY	(2) SMR CODE e USMC	(3) NATIONAL STOCK NUMBER	(4) DESCRIPTION REF NUMBER MFR CODE	(4) USABLE ON CODE	(5) U/M	(6) QTY INC IN	(7) USMC QTY PER EQUIP
							GROUP 05 ENGINE				
28		PAFZZ	PAFZZ	PAFZZ	PAFZZ	2815-01-274-9514	PISTON 112-241 (44940)	ABC	EA 1		2
28		PAFZZ	PAFZZ	PAFZZ	PAFZZ	2815-01-275-5364	PISTON OVS................ 112-0242 (44940)	ABC	EA 1		2
28		KFZZF	KFFZZ	KFFZZ	KFFZZ		PISTON OVS................ 112-0232-01 (44940) PART OF KIT 12-0242 (44940)	ABC	EA I		2
28		KFFZZ	KFFZZ	KFFZ	KFFZZ		PIN, PISTON .... 112-213 (44940) PART OF KIT 112-0242 (44940)	ABC	EA 1		4
28		KFFZZ	KFFZZ	KFFZZ	KFFZZ		RING, RETAINING .... 518-0399 (44940) PART OF KIT 112-0242 (44940)	ABC	EA 2		8
28		PAFZZ	PAFZZ	PAFZZ	PAFZZ	5365-01-2S1-1124	RING SET, PISTON................ 113-0270 (44940)	ABC	EA 1		2
28		PAFZZ	PAFZZ	PAFZZ	PAFZZ	2815-01-274-9239	RING SET, PISTON, OVS 113-0271 (44940)	ABC	EA 1		2
28		KFFZZ	KFFZZ	KFFZZ	KFFZZ	2815-01-274-9240	RING,OIL OVS .... 113-0266-01 (44940) PART OF KIT 113-071 (44940)	ABC	EA 1		2
28		KFFZZ	KFMZZ	KFFZZ	KFFZZ		RING,2ND,OVS .... 113-266-01 (44940) PART OF KIT 113-0271 (44940)	ABC	EA 1		2
28		KFFZ	KFFZZ	KFFZZ	KFFZZ		RING,TOP OVS................ 113-0268O1 (44940) PART OF KIT 113-0271 (44940)	ABC	EA 1		2
28		KFFZZ	KFFZZ	KFFZZ	KFFZZ		RING, RETAINING .... 518-0399 (44940) PART OF KIT 12-0241 (44940)	ABC	EA 1		2
28	1	KFFZZ	KFFZZ	KFFZZ	KFFZZ	5365-01-281-1124	PIN.INPISTON................ 112-0213 (44940) PART OF KIT 112-0241 (44940)	ABC	EA 2		1
28	2	KFFZZ	KFFZZ	KFFZZ	KFFZZ		PISTON .... 112-0232-00 (44940) PART OF KIT 112-0241 (44940)	ABC	EA 1		2
28	3	KFFZZ	KFFZZ	KFFZZ	KFFZZ		RING,TOP COMP................ 113-0268-00 (44940) PART OF KIT 113-0270 (44940)	ABC	EA 1		2
	4								EA 1		

(1) ILLUS-TRA-TION (a) FIG NO.	(b)	(2) SMR CODE a ARMY	b AIR FORCE	d NAVY	e USMC	(3) NATIONAL STOCK NUMBER	(4) DESCRIPTION USABLE ON CODE REF NUMBER MFR CODE	(5) U/M	(6) QTY INC IN	(7) USMC QTY PER EQUIP
							GROUP 05 ENGINE			
28	5	KFFZZ	KFFZZ	KFFZZ	KFFZZ		RING,2ND COMP. ABC 113-2674-00 (44940) PART OF KIT 113-270 (44940)	EA	1	2
28	6	KFFZZ	KFFZZ	KFFZZ	KFFZZ		RING,OIL ABC 113-0266-00 (44940) PART OF KIT 113-0270 (44940)	EA	I	2
28	7	PAFZZ	PAFZZ	PAFZZ	PAFZZ	215-01-274-6813	CONNECTNG ROD ASM ABC 114-0331 (44940)	EA	1	I
28	8	XAFZZ	XAFZZ	XAFZZ	XAFZZ	5310-01-240-1472	NUT ABC C0114031800 (44940)	EA	2	2
28	9	XAFZZ	XAFZ	XAFZZ	XAFZZ	5306-01-228-7458	BOLT ABC 114-0315 (44940)	EA	2	2
28	10	XAFZZ	XAFZZ	XAFZZ	XAFZZ		CONNECTING,ROD ABC 114-0331-01 (44940)	A	1	1
28	11	PAFZZ	PAFZZ	PAFZZ	PAFZZ		BUSHING, SLEEVE ABC 114-0305 (44940)	EA	1	1
28	12	PAFZZ	PAFZZ	PAFZZ	PAFZZ	3120-01-276-3385	BEARING, SLEEVE ABC 114-0279-00 (44940)	EA	2	2
28		PAFZZ	PAFZZ	PAFZZ	PAFZZ	3120-01-277-2403	KIT,PISTON ABC 112-0241 (44940)	EA	1	1
28		KFFZZ	KFFZZ	KFFZ	KFFZZ		RING,RETAIN PISTON ABC 518-0399 (44940)	EA	2	2
28		KFFZZ	KFFZZ	KFFZZ	KFFZZ		PIN, PISTON ABC 112-0213 (44940)	EA	1	1
28		KFFZZ	KFFZZ	KFFZZ	KFFZZ		PISTON ABC 112-O232-00 (44940)	EA	1	1
28		PAFZZ	PAFZZ	PAFZZ	PAFZZ		KIT,PISTON O/S ABC 112-0242 (44940)	EA	1	1
28		KFFZZ	KFFZZ	KFFZZ	KFFZZ		PISTON OVER SIZE ABC 112-0232-01 (44940)	EA	1	1
28		KFFZZ	KFFZZ	KFFZZ	KFFZZ		PIN, PISTON ABC 112-0213 (44940)	EA	1	1
28		KFFZZ	KFFZZ	KFFZZ	KFFZZ		RING,RETAIN PISTON ABC 518-0399 (44940)	EA	2	2
28		PAFZZ	PAFZZ	PAFZZ	PAFZZ		KIT,RING SET ABC 113-0270 (44940)	EA	1	1

(1) ILLUSTRATION (a) FIG NO.	(1) ILLUSTRATION (b)	(2) SMR CODE a ARMY	(2) SMR CODE b AIR FORCE	(2) SMR CODE d NAVY	(2) SMR CODE e USMC	(3) NATIONAL STOCK NUMBER	(4) DESCRIPTION REF NUMBER MFR CODE	(4) USABLE ON CODE	(5) U/M	(6) QTY INC IN	(7) USMC QTY PER EQUIP
							GROUP 05 ENGINE				
28		KFFZZ	KFFZZ	KFFZZ	KFFZZ		RING,TOP COMP................................. 113-0268-00 (44940)	ABC	EA 1		1
28		KFFZZ	KFFZZ	KFFZZ	KFFZZ		RING,2ND COMP................................... 113-0267-00 (44940)	ABC	EA 1		1
28		KFFZZ	KFFZZ	KFFZZ	KFFZZ		RING,OIL... 113-0266-00 (44940)	ABC	EA I		1
28		PAFZZ	PAFZZ	PAFZZ	PAFZZ		KIT,RING OVS...................................... 113-0271 (44940)	ABC	EA 1		1
28		KFFZZ	KFFZZ	KFFZZ	KFFZZ		RING,OIL OVS 113-0266-01 (44940)	ABC	EA 1		1
28		KFFZZ	KFFZZ	KFFZZ	KFFZZ		RING,2ND, OVS 113-0267-01 (44940)	ABC	EA 1		1
28		KFFZZ	KFFZZ	KFFZZ	KFFZZ		TOP RING OVS.................................... 113-0268-01 (44940)	ABC	EA 1		1

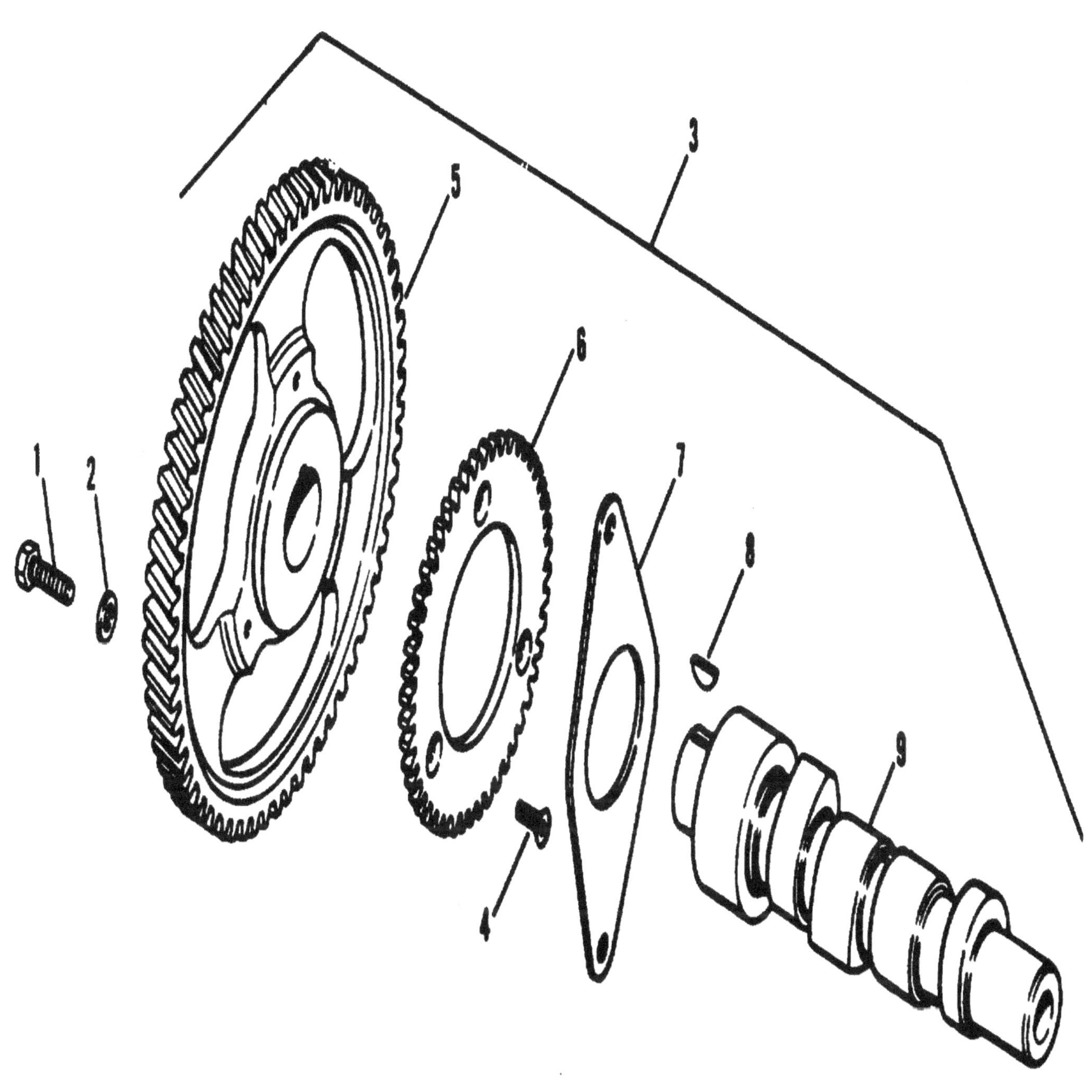

Figure 29. Camshaft, Group 05 (Engine)

(1) ILLUSTRATION (a) FIG NO.	(b)	(2) SMR CODE a ARMY	b AIR FORCE	d NAVY	e USMC	(3) NATIONAL STOCK NUMBER	(4) DESCRIPTION REF NUMBER MFR CODE	USABLE ON CODE	(5) U/M	(6) QTY INC IN	(7) USMC QTY PER EQUIP
							GROUP 05 ENGINE				
29	1	PAFZZ	PAFZZ	PAFZZ	PAFZZ	5305-01-212-3221	SCREW 718-1021 (44940)	ABC	EA 2		2
29	2	PAFZZ	PAFZZ	PAFZZ	PAFZZ	5310-01-275-3318	WASHER, FLAT 740-1004 (44940)	ABC	EA 2		2
29	3	AFFZZ	AFFZZ	AFFZZ	AFFZZ		CAMSHAFT ASSEM 105-0609 (44940)	ABC			I
29	4	PAFZZ	PAFZZ	PAFZZ	PAFZZ		SCREW, MACHINE 802-2505 (44940)	ABC	EA I		3
29	5	PAFZZ	PAFZZ	PAFZZ	PAFZZ	5305-ol1-276-993	GEAR, SPUR 105-0608 (44940)	ABC	EA 3		I
29	6	PAFZZ	PAFZZ	PAFZZ	PAFZZ	3020-01-274-9402	GEAR, SPUR 150-2013 (44940)	ABC	EA I		I
29	7	PAFZZ	PAFZZ	PAFZZ	PAFZZ	3020-01-274-9403	PLATE, RETAINING 105-0587 (44940)	ABC	EA I		1
29	8	PAFZZ	PAFZZ	PAFZZ	PAFZZ		KEY, MACHINE 515-1 (44940)	ABC			2
29	9	PAFZZ	PAFZZ	PAFZZ	PAFZZ	3040-01-27-9617	CAMSHAFT 105-0599 (44940)	ABC	EA I		1
						5315-00-652-8675			EA 1		
						215-01-275-			EA I		

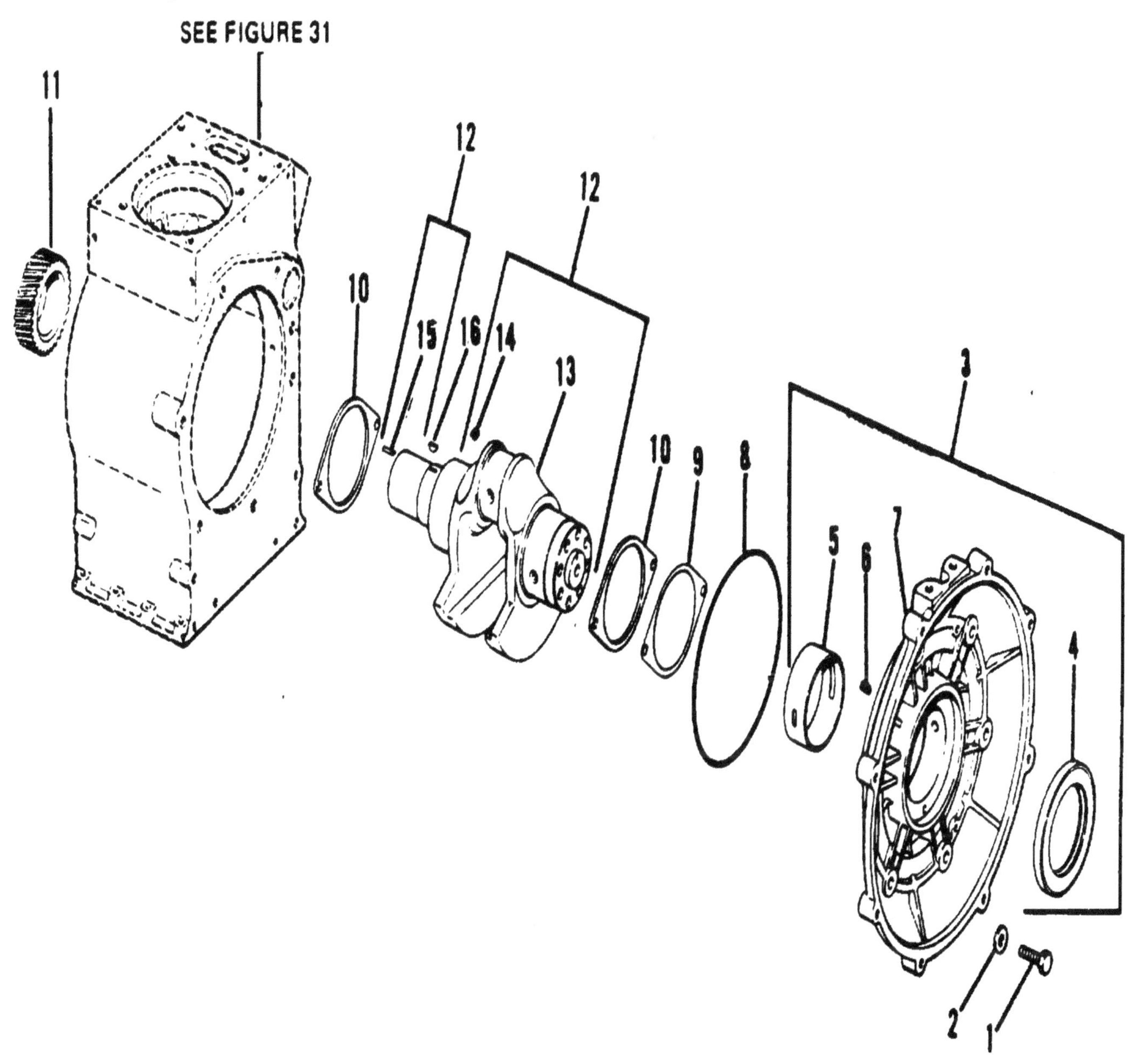

Figure 30. Crankshaft, Group 05 (Engine)

(1) ILLUSTRATION (a) FIG NO.	(b)	(2) SMR CODE a ARMY	b AIR FORCE	d NAVY	e USMC	(3) NATIONAL STOCK NUMBER	(4) DESCRIPTION USABLE ON CODE REF NUMBER MFR CODE	(5) U/M	(6) QTY INC IN	(7) USMC QTY PER EQUIP
							GROUP 05 ENGINE			
30 PAF7Z	1		PAFZZ	PAFZZ	PAFZZ	5306-01-275-3243	BOLT, MACHINE.................................. ARC 718-1055 (44940)	EA	6	6
			PAFZZ	PAFZZ	PAFZZ	5310-01-275-3321	WASHER,FLAT .... ABC 740-1008 (44940) C07400100800 (15434)	EA	6	6
30 PAFZZ	2									
			AFFZZ	AFFZZ	AFFZZ		ADAPTER ASSEM,GEN........................ ABC 231-0272 (44940)	EA	1	1
30	3	AFFZZ								
			PAFZZ	PAFZZ	PAFZZ	5330-01-275-1956	SEAL, PLAIN .... ABC 509-0204 (44940) PART OF KIT 16840184 (44940)	EA	1	2
30 PAFZZ	4									
			PAFZZ	PAFZZ	PAFZZ	3120-01-276-3:75	BEARING, SLEEVE.............................. ABC 101-0473-00 (44940)	EA	1	2
30 PAFZZ	5		PAFZZ	PAFZZ	PAFZZ	5315-01-275-3451	PIN, SPRING.. ABC 516-2003 (44940)	EA	2	4
30 PAFZZ	6		XBFZZ	XBF7Z	XBFZZ		HOUSING,END BELL........................... ABC 231-0271 (44940)	EA	I	1
			PAFZZ	PAFZZ	PAFZZ	5330-01-275-1961	PACKING, PREFORMED..................... ABC 509-0163 (44940) PART OF KIT 168-O184 (44940)	EA	1	2
30 XBFZZ	7									
			PAFZZ	PAFZZ	PAFZZ	5365-01-275-644	SHIM W/O HOLE.................................. ABC 104-1565-01 (44940)	EA	1	1
30 PAFZZ	8									
			PAFZZ	PAFZZ	PAFZZ	5365-01-275-68,5	SHIM,THRUST W/HOLE .... ABC 104-1565-02 (44940)	EA	1	1
30 PAFZZ	9									
			PAHZZ	PAHZZ	PAHZZ	3120-01-276-3389	BEARING, WASHER............................ ABC 104-1564-02 (44940)	EA	2	2
30 PAFZZ	9		PAHZZ	PAHZZ	PAHZZ	3020-01-274-9404	GEAR, SPUR.. ABC	EA	1	1
			PAHHH	PAHHH	PAHHH	2815-01-275-1711	CRANKSHAFT, ENGINE...................... ABC 104-1646 (44940)	EA	1	1
30 PAHZZ	10									
			XAHZZ	XAHZZ	XAHZZ		CRANKSAFT.. ABC 104-1631 (44940)	EA	1	1
30 PAHZZ 104-1636 (44940)	11		PAHZZ	PAHZZ	PAHZZ	4730-00- 808-6814	PLUG... ABC WW-P-471 (81346) 1/8HHP-S (98441)	EA	1	4
			PAHZZ	PAH'ZZ	PAHZZ	5315-01-275-6994	PIN, STRAIGHT.................................... ABC 775-0035 (44940)	EA	1	1
30 PAHHH	12									
			PAHZZ	PAHZZ	PAHZZ	5315- 00-652-8675	KEY,MACHINE .... ABC 515-1 (44940)	EA	1	1
30 XAHZZ	13									
30 PAHZZ	14									

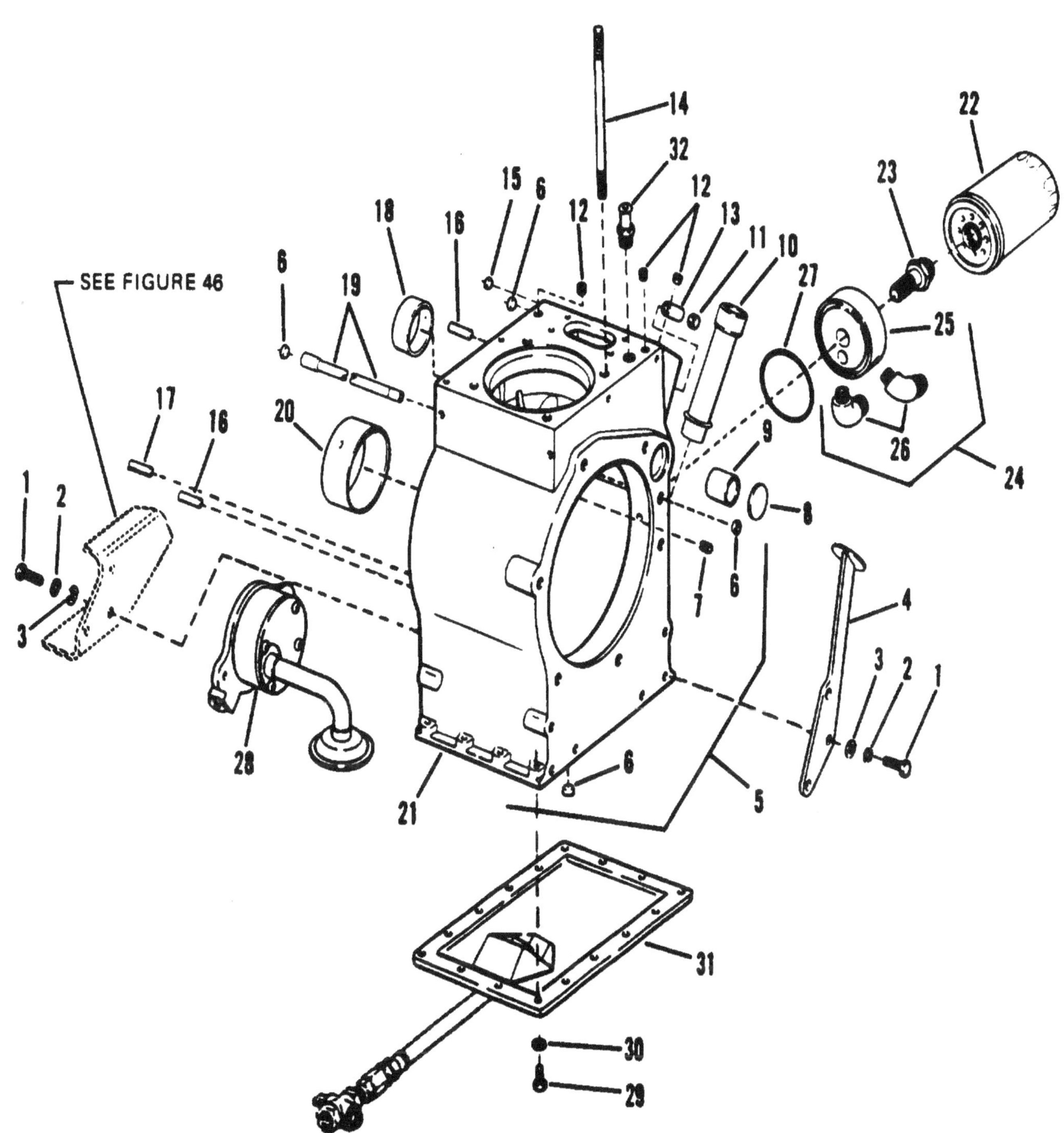

Figure 31. Crankcase, Group 05 (Engine)

(1) ILLUSTRATION (a) FIG NO.	(b)	(2) SMR CODE a ARMY	b AIR FORCE	d NAVY	e USMC	(3) NATIONAL STOCK NUMBER	(4) DESCRIPTION REF NUMBER MFR CODE	USABLE ON CODE	(5) U/M	(6) QTY INC IN	(7) USMC QTY PER EQUIP
							GROUP 05 ENGINE				
31	1	PAFZZ	PAFZZ	PAFZZ	PAFZZ	530-01-174-8739	BOLT, MACHINE B18231A08025N (80204)	ABC	EA	16	16
31	2	PAFZZ	PAFZZ	PAFZZ	PAFZZ	5310-4On-9566	WASH,LOCK MS35338-45 (96906) 850-1045 (44940)	ABC	EA	16	16
31	3	PAFZZ	PAFZZ	PAFZZ	PAFZZ	5310X4)14219	WASHSER,FLAT MS27183-12 (96906)	ABC	EA	16	16
31	4	PAFZZ	PAFZZ	PAFZZ	PAFZZ	1560-01-28e0-708	SUPPORT, STRUCTURAL 403-2371 (44940) 84-130T (30554)	ABC	kA	1	1
31	5	PAHZZ	PAHZZ	PAHZZ	PAHZZ	2815-01-27-3515	ENGINE BLOCK ASSY 101-0756 (44940) 101-0771 (44940)	ABC	EA 1	1	
31	6	XBHZZ	XBHZZ	XBHZZ	XBHZZ		PLUG, CUP 517-0218 (44940)	ABC	EA 4	4	
31	7	XBHZZ	XBHZZ	XBHZZ	XBHZZ	5315-01-275-451	PIN, SPRING 51-2003 (44940)	ABC	EA 2	2	
31	8	XBHZZ	XBHZZ	XBHZZ	XBHZZ		PLUG, WELCH 517-0219 (44940)	ABC	EA 1	1	
31	9	PAHZZ	PAHZZ	PAHZZ	PAHZZ	3120-01-277-4435	BEARING, SLEEVE 101-0603 (44940)	ABC	EA 1	1	
31	10	XBHZZ	XBHZZ	XBHZZ	XBHZZ		TUBE,OIL FILL 123-1684 (44940)	ABC	EA 1	1	
31	11	PAHZZ	PAH2Z	PAHZZ	PAHZ	5330-01-275-6832	SEAL, PLAIN 509-0211 (44940) PART OF KIT 1684 (44940)	ABC	EA 2	1	
31	12	PAHZZ	PAHZZ	PAHZZ	PAHZZ	4730-0-808-6814	PLUG WW-P-471 (81348) 1/8HHP-S (98441)	ABC	EA	3	3
31	13	PAHZZ	PAHZ	PAHZZ	PAHZZ	3120-01-291-9234	BEARING, SLEEVE 510-0151 (44940)	ABC	EA 1	1	
31	14	PAHZZ	PAHZZ	PAHZZ	PAHZZ	5307-01-276-7534	STUD 520-2408 4940)	ABC	EA 4	4	
31	15	XBHZZ	XBHZZ	XBHZZ	XBHZZ		PLUG,CUP 517-0217 (44940)	ABC	EA 1	1	
31	16	XBHZZ	XBHZZ	XBHZZ	XBHZZ		PIN,PARALLEL 775-0075 (44940)	ABC	EA 2	2	
31	17	XBHZZ	XBHZZ	XBHZZ	XBHZZ		PIN, PARALLEL 775-O73 (44940)	ABC	RA	1	1

(1) ILLUSTRATION (a) FIG NO.	(1) ILLUSTRATION (b) ITEM NO.	(2) SMR CODE a ARMY	(2) SMR CODE b AIR FORCE	(2) SMR CODE d NAVY	(2) SMR CODE e USMC	(3) NATIONAL STOCK NUMBER	(4) DESCRIPTION REF NUMBER MFR CODE	(4) USABLE ON CODE	(5) U/M	(6) QTY INC IN	(7) USMC QTY PER EQUIP
							GROUP 05 ENGINE				
31	18	PAHZZ	PAHZZ	PAHZZ	PAHZZ	3120-01-277-8898	BEARING, SLEEVE 101-0572 (44940)	ABC	EA	1	1
31	19	XBHZZ	XBHZZ	XBHZZ	XBHZZ		TUBE,OILTRANSFER 120-1123 (44940)	ABC	EA	1	1
31	20	PAHZZ	PAHZZ	PAHZZ	PAHZZ	3120-01-276-3375	BEARING, SLEEVE 101-0473-00 (44940)	ABC	EA	1	1
31	21	XAHZZ	XAHZZ	XAHZZ	XAHZZ		CRANKCASE 101-0749 (44940)	ABC	EA	1	1
31	22	PAOZZ	PAOZZ	PAOZZ	PAOZZ	2940-01-275-4285	FILTER, ELEMENT 122-0602-01 (44940) LF3525 (33457)	ABC	EA	1	1
31	23	PAFZZ	PAFZZ	PAFZZ	PAFZZ		ADAPTER OIL FILTER 122-0682 (44940)	ABC	EA	1	1
31	24	AFFZZ	AFFZZ	AFFZZ	AFFZZ	2940-01-275-9157	ADAPTER ASSY,OIL 102-1324 (44940)	ABC	EA	1	1
31	25	PAFZZ	PAFZZ	PAFZZ	PAFZZ		ADAPTER, OIL COOLER 102-1323 (44940)	ABC	EA	1	1
31	26	PAFZZ	PAFZZ	PAFZZ	PAFZZ	2930-01-276-5970	ELBOW 502-1015 (44940) MS51504AS-4 (96906)	ABC	EA	2	2
31	27	PAFZZ	PAFZZ	PAFZZ	PAFZZ	4730-00-844-3308	PACKING, PREFORMED 509-O228 (44940) PART OF KIT 168-	ABC	EA	1	2
31	28	PAFZZ	PAFZZ	PAFZZ	PAFZZ	5330-01-276-1681	PUMP, OIL 120-1131 (44940)	ABC	EA	1	1
31	29	PAFZZ	PAFZZ	PAFZZ	PAFZZ		BOLT, MACHINE 718-1018 (44940)	ABC	EA	16	16
31	30	PAFZZ	PAFZZ	PAFZZ	PAFZZ	2815-01-274-9361	WASHER, FLAT 740-1004 (44940)	ABC	EA	16	16
31	31	PAFZZ	PAFZZ	PAFZZ	PAFZZ	5306-01-275-6000	PAN,OIL 102-1259 (44940)	ABC	EA	1	1
31	32	PAFZZ	PAFZZ	PAFZZ	PAFZZ	5310-01-275-3318	ADAPTER, STRAIGBT 502-0979 (44940) 68HB-8-4 (93061)	ABC	EA	1	1
31		PAFZZ	PAFZZ	PAFZ	PAFZZ	2815-01-274-9354	KIT,GASKET SET 168-0184 (44940)	ABC	EA	1	1
31		PAFZZ	PAFZZ	PAFZZ	PAFZZ	4730-01-275-4180	GASKET 149-2060 (44940)	ABC	EA	1	1
31		PAFZZ	PAFZZ	PAFZZ	PAFZZ	5330-01-283-4297	BUSHING, NONMETALLIC 502-0953 (44940)	ABC	EA	2	2

(1) ILLUSTRATION (a) FIG NO.	(b) ITEM NO.	(2) SMR CODE a ARMY	b AIR FORCE	d NAVY	e USMC	(3) NATIONAL STOCK NUMBER	(4) DESCRIPTION REF NUMBER MFR CODE	USABLE ON CODE	(5) U/M	(6) QTY INC IN	(7) USMC QTY PER EQUIP
							GROUP 05 ENGINE				
31		PAFZZ	PAFZZ	PAFZZ	PAFZZ	5330-01-276-7501	GASKET... 150-2133 (44940)	ABC	EA 1	1	
31		PAFZZ	PAMZ	PAFZZ	PAFZZ	5330-01-275-3358	GASKET... 154-2322 (44940)	ABC	EA 1	1	
31		PAFZZ	PAFZZ	PAFZZ	PAFZZ	5310-01-275-3321	WASHER, FLAT 147-0430 (44940)	ABC	A 1	1	
31		PAFZZ	PAFZZ	PAFZZ	PAFZZ	5330-01-275-3359	GASKET... 154-2342 (44940)	ABC	EA 1	1	
31		PAFZZ	PAFZZ	PAFZZ	PAFZZ	5330-01-275-3360	GASKET... 115-0400..(44940)	ABC	EA 1	1	
31		PAFZZ	PAFZZ	PAFZZ	PAFZZ	2940-01-274-456	FILTER, ELEMENT................................ 123-1707 (44940)	ABC	EA 1	1	
31		PAFZZ	PAFZZ	PAFZZ	PAFZZ	5330-01-275-3361	GASKET... 115-0351 (44940)	ABC	EA 1	1	
31		PAFZZ	PAFZZ	PAFZZ	PAFZZ	2590-01-274-9241	RING, WIPER... 509-0221 (44940)	ABC	EA 2	2	
31		PAFZZ	PAFZZ	PAFZZ	PAFZZ	5330-01-275-3338	PACKING, PREFORMED...................... 115-0268 (44940)	ABC	EA 4	4	
31		PAFZZ	PAFZZ	PAFZZ	PAFZZ	5330-01-276-2290	GASKET... 115-0403 (44940)	ABC	A 1	1	
31		PAFZZ	PAFZZ	PAFZZ	PAFZZ	5330-01-275-5013	SEAL, PLAIN, ENCASED 509-0166 (44940)	ABC	EA 1	1	
31		PAFZZ	PAFZZ	PAFZZ	PAFZZ	5330-01-275-5014	GASKET... 103-0782 (44940)	ABC	EA 1	1	
31		PAFZZ	PAFZZ	PAFZZ	PAFZZ	5330-01-275-6838	GASKET... 103-0783 (44940)	ABC	EA 1	1	
31		PAFZZ	PAFZZ	PAMZZ	PAFZZ	5330-01-275-1956	SEAL, PLAIN, ENCASED 509-0204 (44940)	ABC	EA 1	1	
31		PAFZZ	PAFZZ	PAFZZ	PAFZZ	5330-01-275-1961	PACKING, PREFORMED...................... 509-0163 (44940)	ABC	EA 1	1	
31		PAFZZ	PAFZZ	PAFZZ	PAFZZ	5330-01-275-6832	SEAL, PLAIN, ENCASED 509-0211 (44940)	ABC	EA 1	1	
31		PAFZZ	PAFZZ	PAFZZ	PAFZZ	5330-01-276-1681	PACKING, PREFORMED..................... 509-0228 (44940)	ABC	EA	1	

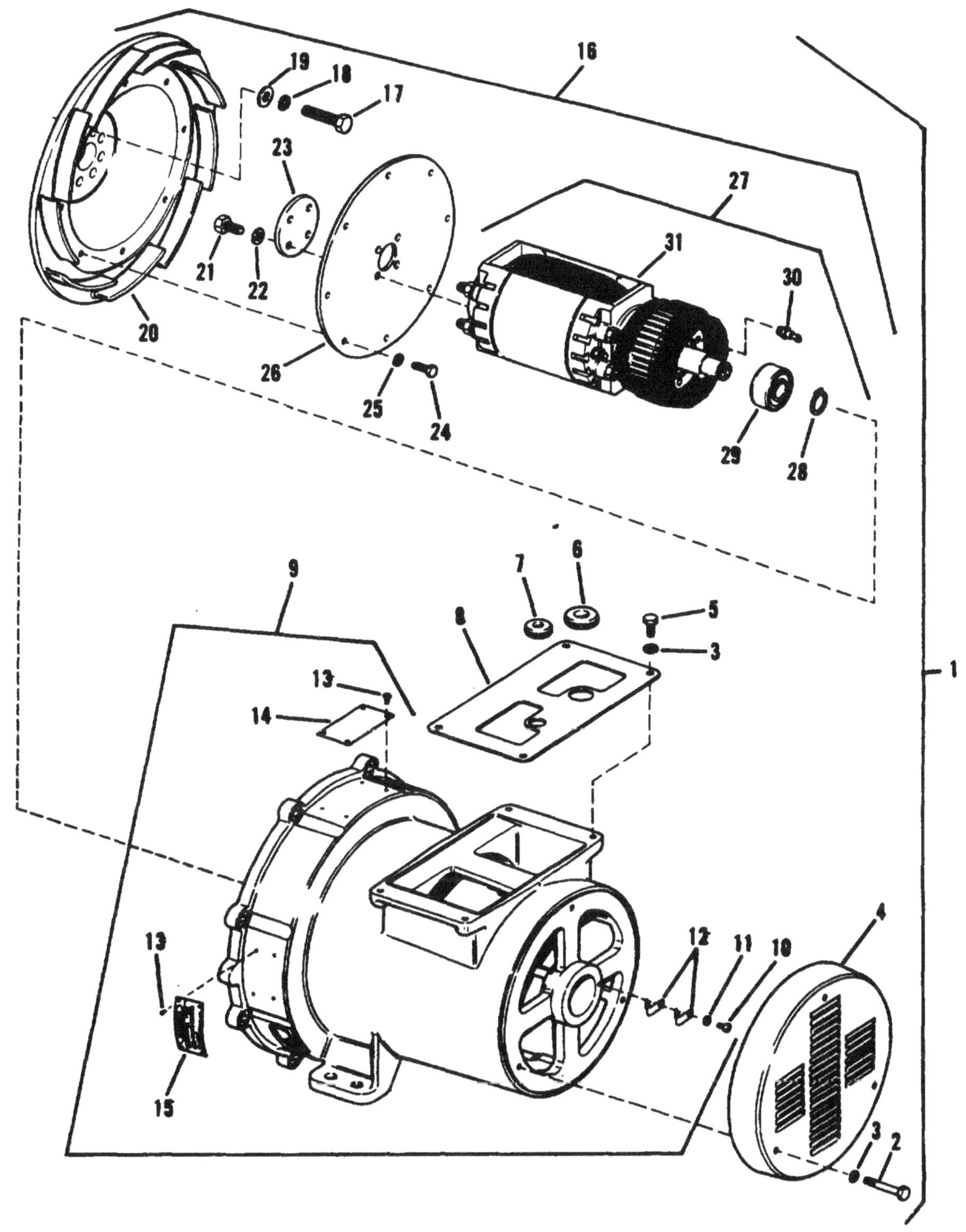

Figure 32. Generator Assembly (28 VDC, 60 Hz), Group 03 (Electrical Power Generation System)

(1) ILLUS-TRA-TION (a) FIG NO.	(b)	(2) SMR CODE a ARMY	b AIR FORCE	d NAVY	e USMC	(3) NATIONAL STOCK NUMBER	(4) DESCRIPTION REF NUMBER MFR CODE	USABLE ON CODE	(5) U/M	(6) QTY INC IN	(7) USMC QTY PER EQUIP
							GROUP 03 ELECIRICAL POWER GENERATION SYSTEM				
32	1	PAFFF	PAFFF	PAFFF	PAFFF	6115-01-289-1081	GENERATOR 200-0999 (449) 84-13135 (30554)	A	EA	1	1
32	1	PAFFF	PAFFF	PAFFF	PAFFF		GNERATOR 28VDC 200-1000 (4490) 84-13136 (30554)	C	EA	1	1
32	2	PAOZZ	PAOZZ	PAOZZ	PAOZZ	5305-00-71-2510	SCREW,CAP MS90728-13 (96906)	AC	EA	3	6
32	3	PAOZZ	PAOZZ	PAOZZ	PAOZZ	5310-X-582-5965	WASH,LOCK MS35338-44 (96906) 850-1040 (44940)	AC	EA	7	22
32	4	XBOZZ	XBOZZ	XBOZZ	XBOZZ		COVER,HOUSING END 234-0805 (44940) 84-13123 (30554)	AC	EA	1	2
32	5	PAFZZ	PAFZZ	PAFZZ	PAFZZ		SCREW,CAP MS90725-3 (96906)	AC	EA	4	8
32	6	PAFZZ	PAFZZ	PAFZZ	PAFZZ	530540-0680500	GROMMET MS35489-48 (96906)	AC	EA	1	2
32	7	PAFZZ	PAFZZ	PAFZZ	PAFZZ	5325-00-174-9332	GROMMET 205K1114 (94990) 2860 (70485)	AC	EA	1	4
32	8	XBFZZ	XBFZZ	XBFZZ	XBFZZ	532510-185-003	COVER 234-8020 (44940) 84-13126 (3O554)	AC	EA	1	2
32	9	PAFZZ	PAFZZ	PAFZZ	PAFZZ		STATOR, GENERATOR 13213E4168 (97403) D15751 (07860)	A	EA	1	1
32	9	PAFZZ	PAFZZ	PAFZZ	PAFZZ	611540-758-9239	STATOR ASSEMBLY 13213E4107 (97403)	C	EA	1	1
32	10	PAFZZ	PAFZZ	PAFZZ	PAFZZ		SCREW, MACHINE MS35206-243 (96906)	AC	EA	1	2
32	11	PAFZZ	PAFZZ	PAFZZ	PAFZZ	6115 -0-997-9769	WASHER, LOCK MS35338-42 (96906)	AC	EA	1	2
32	12	PAFZZ	PAFZZ	PAFZZ	PAFZZ	5305-00-9446191	CLIP, SPRING 13213E4228 (97403) 232-2015 (44940)	AC	EA	2	2
						5310-04M5-3299					

(1) ILLUSTRATION (a) FIG NO.	(1) (b)	(2) SMR CODE a ARMY	(2) b AIR FORCE	(2) d NAVY	(2) e USMC	(3) NATIONAL STOCK NUMBER	(4) DESCRIPTION USABLE ON CODE / REF NUMBER MFR CODE	(5) U/M	(6) QTY INC IN	(7) USMC QTY PER EQUIP
							GROUP 03 ELECIRICAL POWER GENERATION SYSTEM			
32 PAFZ	13		PAF PAZ	PAFUZ	PAFZZ	530540-175-3230	SCREW,DRIVE ....AC MS21318-14 (96906)	EA	12	30
			MDFZZ	MDFZZ	MDFZZ		PLATE, IDENTIFICATION .... A 13214E6027 (97403)	EA	1	1
32 MDFZZ	14					9905-0143S-7439				
			MDFZZ	MDFZZ	MDFZZ		PLATE, IDENTIFICATION .... C 13214E9580 (97403)	EA	1	2
32 MDFZZ	14					9905014-66-3081				
			MDFZZ	MDFZZ	MDFZZ		PLATE,WARNING ....AC 099-2343 (44940) 84-13210 (30554)	EA	1	2
32 MDFZZ	15									
			AFFF	AFFF	AFFF		ROTOR ASSY ....AC 201-1557 (44940) 84-13125 (30554)	EA	2	2
						6115-01-292-6970				
32 AFFF	16		PAFZZ	PAFZZ	PAFZZ		BOLT, MACHINE....AC 000931 010355 (64678)	EA	8	16
			PAFZZ	PAFZZ	PAFZZ	5306-01-238-3172	WASHER, LOCK ....AC MS35338-47 (96906) 850-1055 (44940)	EA	8	16
32 PAFZZ	17									
			PAFZZ	PAFZZ	PAFZZ	5310-00-209-965	WASHER, FLAT ....AC MS27183-15 (96906)	EA	1	9
32 PAFZZ	18									
			XBFZZ	XBFZZ	XBFZZ		FAN, ROTOR....AC 205-0079 (44940) 84-13122 (30554)	EA	1	2
32 PAFZZ	19					5310)0-809-4061				
			PAFZZ	PAFZZ	PAFZZ		BOLT, MACHINE....AC MS90728-32 (96906)	EA	4	4
32 XBFZZ	20		PAFZZ	PAFZZ	PAFZZ		WASHER,LOCK ....AC MS35338-45 (96906) 850-1045 (44940)	EA	8	a
			XBFZZ	XBFZZ	XBFZZ	530600-226-4825	PLATE,WASHER....AC 232-3228 (44940) 84-13127 (30554)	EA	1	2
32 PAFZZ	21									
			PAFZZ	PAFZZ	PAFZZ	5310-00-407-9566	SCREW,CAP, HEXAGON ....AC MS90725-6 (96906) 800-1005 (44940)	EA	8	16
32 PAFZZ	22									
			PAFZZ	PAFZZ	PAFZZ		WASHER,LOCK ....AC MS35333-40 (9606)	EA	8	8
32 XBFZZ	23									
						53054-0681502				
32 PAFZZ	24									
32 PAFZZ	25									

(1) ILLUS-TRA-TION (a) FIG NO.	(1) (b)	(2) SMR CODE a ARMY	b AIR FORCE	d NAVY	e USMC	(3) NATIONAL STOCK NUMBER	(4) DESCRIPTION REF NUMBER MFR CODE	USABLE ON CODE	(5) U/M	(6) QTY INC IN	(7) USMC QTY PER EQUIP
							GROUP 03 ELECIRICAL POWER GENERATION SYSTEM				
32	26	XBZZ	XBFZZ	XBPFZ	XBFZ	6115L-9-2410	PLATE, COUPLING 132134097 (97403) 2322009 (44940)	AC	EA	1	2
32	27	PAFZZ	PAFZZ	PAFZZ	PAFZZ	2920-01-276-6903	ROTOR ASSY 201-3417 (44940)	AC	EA 1	1	
32	28	PAFZZ	PAFZZ	PAFZZ	PAFZZ	5365-00-804-7659	RING, RETAINING MS16624-1078 (96906) 518-0279 (44940)	AC	EA 2	1	
32	29	PAFZZ	PAFZZ	PAFZZ	PAFZZ	3110-00-158-8243	BEARING 13214E6024 (97403) 510-0097 (44940)	AC	EA 1	1	
32	30	PAFZZ	PAFZZ	PAFZZ	PAFZZ	5961-00-724-5970	SEMICODUOCTOR DEVICE JAN1N1204A (81349)	AC	EA	3	6
32	31	XAFZZ	XAFZZ	XAFZZ	XAFZZ		ROTOR 201-1555 (44940) 13213E4222 (97403)	AC	EA 1	1	

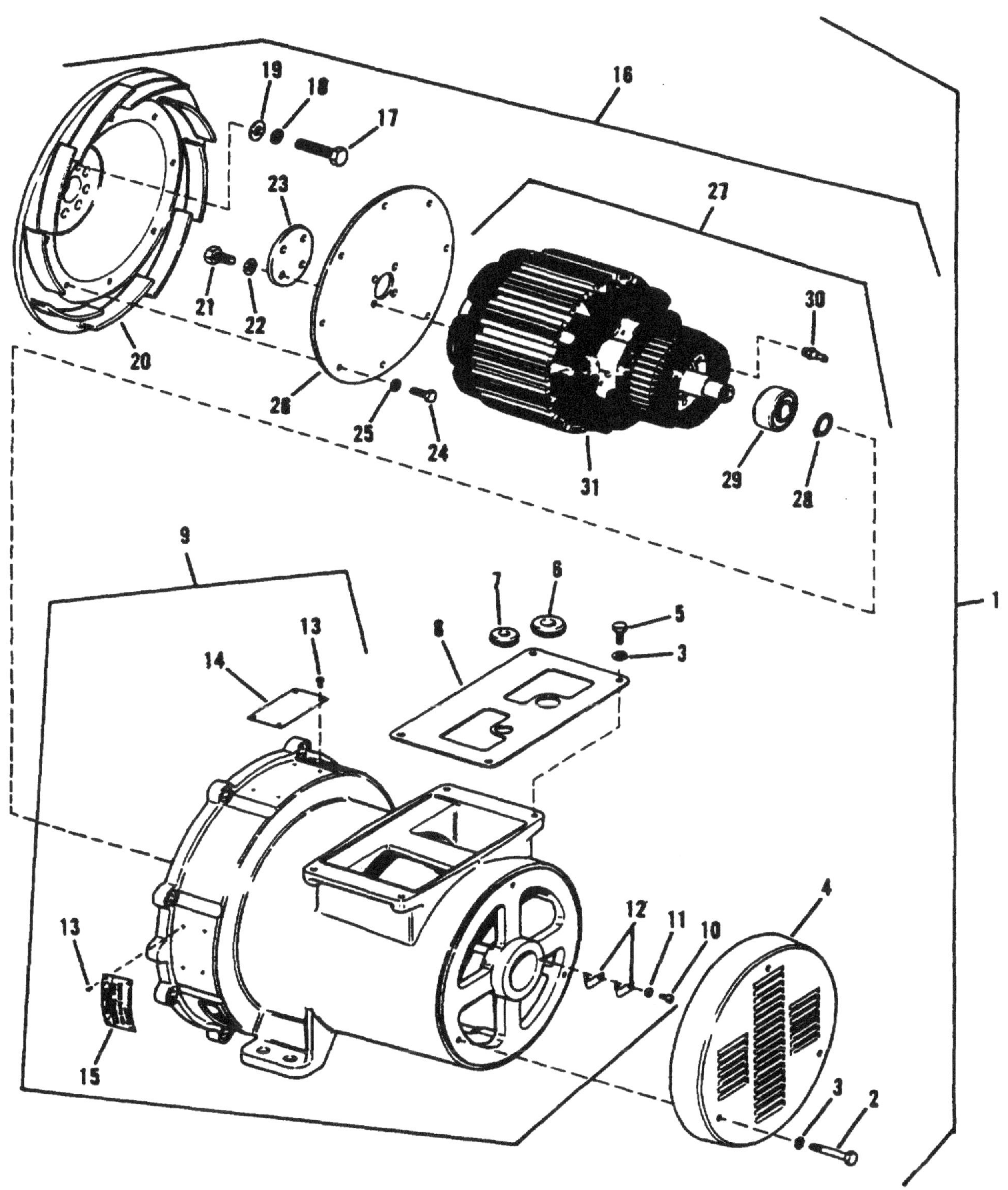

Figure 33. Generator Assembly (400 Hz), Group 03 (Electrical Power Generation System)

(1) ILLUS-TRA-TION (a) FIG NO.	(b)	(2) SMR CODE a ARMY	b AIR FORCE	d NAVY	e USMC	(3) NATIONAL STOCK NUMBER	(4) DESCRIPTION REF NUMBER MFR CODE / USABLE ON CODE	(5) U/M	(6) QTY INC IN	(7) USMC QTY PER EQUIP
							GROUP 03 ELECIRICAL POWER GENERATION SYSTEM			
33	1	PAFFF	PAFFF	PAFFF	PAFFF	6115-01-295-8302	GENERATOR.......... B 200-1023 (44940) 84-13134 (30554)	EA 1	1	
33	2	PAFZZ	PAFZZ	PAFZZ	PAFZZ	530540-71-2510	SCREW, CAP HH.......... ABC MS90728-13 (96906)	EA 3	3	
33	3	PAFZZ	PAFZZ	PAFZZ	PAFZZ	531 50-52-5965	WASHER, LOCK .... ABC MS35338-44 (96906) 850-1040 (44940)	EA	7	7
33	4	PAFZZ	PAFZZ	PAFZZ	PAPZZ		COVER,HOUSING END.......... ABC 234-O805 (44940) 84-13123 (30554)	EA 1	1	
33	5	PAFZZ	PAFZZ	PAFZZ	PAFZZ	5305-0468400	SCREW, CAP HH.......... ABC MS90725-3 (96906)			
33	6	PAFZZ	PAFZZ	PAFZZ	PAFZZ		GROMMET .... ABC MS35489-48 (96906)	EA	4	4
						5325-00174-9332				
33	7	PAFZZ	PAFZZ	PAFZZ	PAFZZ		GROMMET.......... ABC 2860 (70485) 205K1114 (94990)	EA 1	1	
						532540-185-0003				
33	8	PAFZZ	PAFZZ	PAFZZ	PAFZZ		COVER, GENERATOR.......... ABC 234-0820 (44940) 84-13126 (30554)	EA 1	1	
1										
33	9	PAFZZ	PAFZZ	PAFZZ	PAFZZ		HOUSING, ELECTRICAL .... B 13213E4101 (97403)	EA 1	1	
33	10	PAFZZ	PAFZZ	PAFZZ	PAFZZ	6150-00-949-0604	SCREW, MACHINE.......... ABC MS35206-243 (96906)			
								EA 1	1	
33	11	PAFZZ	PAFZZ	PAFZZ	PAFZZ	5305-00-984-6191	WASHER, LOCK .... ABC MS35338-42 (96906)			
33	12	PAFZZ	PAFZZ	PAFZZ	PAFZZ		CLIP, SPRING.......... ABC 232-2015 (44940) 13213E4228 (97403)	EA 1	1	
						53104045-3299				
33	13	PAFZZ	PAFZZ	PAFZZ	PAFZZ	5340-00-930-3386	SCREW,DRIVE .... ABC MS21318-14 (96906)	EA	1	1
								EA	1	1
33	14	MDFZZ	MDFZZ	MDFZZ	MDFZZ		PLATE, IDENTIFICATION .... B 13214E9580 (97403)			
						5305--175-3230		EA	12	12
						9905-10-066-		EA 1	1	

(1) ILLUSTRATION (a) FIG NO.	(1) (b)	(2) SMR CODE a ARMY	b AIR FORCE	d NAVY	e USMC	(3) NATIONAL STOCK NUMBER	(4) DESCRIPTION REF NUMBER MFR CODE	USABLE ON CODE	(5) U/M	(6) QTY INC IN	(7) USMC QTY PER EQUIP
							GROUP 03 ELECIRICAL POWER GENERATION SYSTEM				
33	15	MDFZZ	MDFZZ	MDFZZ	MDFZZ		PLATE, WARMNG 099-2343 (44940) 84-13210 (30544)	ABC	EA	1	1
33	16	PAFFF	PAFFF	PAFFF	PAFFF	6115-01-298-1566	ROTOR ASSY 201-1673 (44940) 84-13124 (30554)	B	EA	1	1
33	17	PAFZZ	PAFZZ	PAFZZ	PAFZZ	5306-01-238-3172	BOLT, MACHINE 000931 010355 (64678)	ABC	EA	8	8
33	18	PAFZZ	PAFZZ	PAFZZ	PAFZZ	531040-209-965	WASHER, LOCK MS35338-47 (96906) 850-1055 (44940)	ABC	EA	8	8
33	19	PAFZZ	PAFZZ	PAFZZ	PAFZZ	5310-0-8094061	WASHER, FLAT MS27183-15 (9696)	ABC	EA	8	8
33	20	XBFZZ	XBFZZ	XBFZZ	XBFZZ		FAN ROTOR 205-4079 (44940) 84-13122 (30554)	ABC	EA	1	1
33	21	PAFZZ	PAFZZ	PAFZZ	PAFZZ	5306-00-2264825	BOLT, MACHINE MS90728-32 (96906)	ABC	EA	4	4
33	22	PAFZZ	PAFZZ	PAFZZ	PAFZZ	5310-407-9566	WASHER,LOCK MS35338-45 (96906) 850-1045 (44940)	ABC	EA	1	1
33	23	XBFZZ	XBFZZ	XBFZZ	XBFZZ		PLATE,WASHER 232-3228 (44940) 84-13127 (30554)	ABC	EA	1	1
33	24	PAFZZ PAFZZ	PAFZZ	PAFZZ		5305-MM-502	SCREW, CAP, HEXAGON MS90725-6 (96906) 800-1005 (44940)	ABC	EA	8	8
33	25	PAFZZ	PAFZZ	PAFZZ	PAFZZ	531040-582-5965	WASHER,LOCK MS35338-44 (96906) 850-1040 (44940)	ABC	EA	8	8
33	26	XBFZZ	XBFZZ	XBFZZ	XBFZZ	6115-00-859-2410	PLATE,COUPLNG 13213E4097 (97403) 232-2009 (44940)	ABC	EA	1	1
33	27	PAFZZ	PAFZZ	PAFZZ	PAFZZ	6115-01-283-0414	ROTOR ASSY 201-3418 (44940)	B	EA	1	1
33	28	PAFZZ	PAFZZ	PAFZZ	PAFZZ	5365-00-804-7653	RING, RETANNG MS16624-1078 (96906) 518-0279 (44940)	ABC	EA	1	1

(1) ILLUSTRATION (a) FIG NO.	(1) ILLUSTRATION (b)	(2) SMR CODE a ARMY	(2) SMR CODE b AIR FORCE	(2) SMR CODE d NAVY	(2) SMR CODE e USMC	(3) NATIONAL STOCK NUMBER	(4) DESCRIPTION REF NUMBER MFR CODE	(4) USABLE ON CODE	(5) U/M	(6) QTY INC IN	(7) USMC QTY PER EQUIP
							GROUP 03 ELECIRICAL POWER GENERATION SYSTEM				
33	29	PAFZZ	PAFZZ	PAFZ	PAFZZ	3110-00-158-8243	BEARING 510-0097 (44940) 13214E6024 (97403)	ABC	EA 1	1	
33	30	PAFZZ	PAFZZ	PAFZZ	PAFZZ	59610-724-5970	SEMICONDUCTOR............................... JAN1N1204A (81349)	ABC	EA 3	3	
33	31	XAFZZ	XAFZZ	XAFZZ	XAFZZ		ROTOR... 201-1674 (44940) 13213E4170 (97403)	B	EA 1	1	

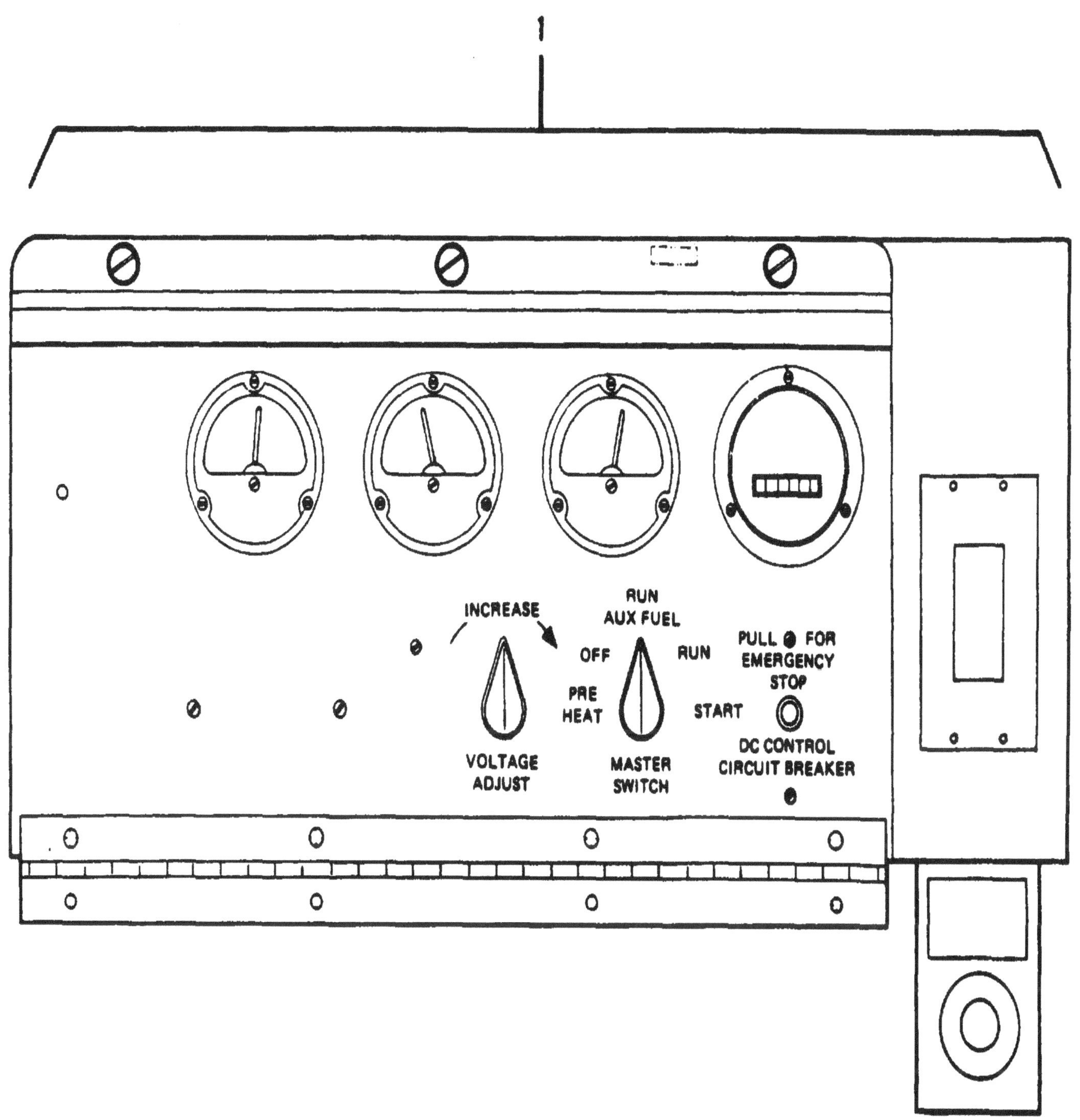

Figure 34. Generator Control 28 VDC, Group 08 (Generator Control and Instruments)

(1) ILLUSTRATION FIG NO	(2) ITEM NO	(3) SMR CODE ARMY	AIR FORCE	NAVY	USMC	(4) NATIONAL STOCK NUMBER	(5) DESCRIPTIION REF NUMBER MFR CODE	(6) USABLE ON CODE	(6) U/M	QTY INC IN UNITS	(7) QTY PER EQUIP
							GROUP 08 GENERATOR CONTROLS AND INSTRUMENTS				
34	1	XBOOO	XBOOO	XBOOO	XBOOO		CONTROL,GEN 28 VDC 300-2962 (44940) 84-13087 (30554)	C	EA	1	1

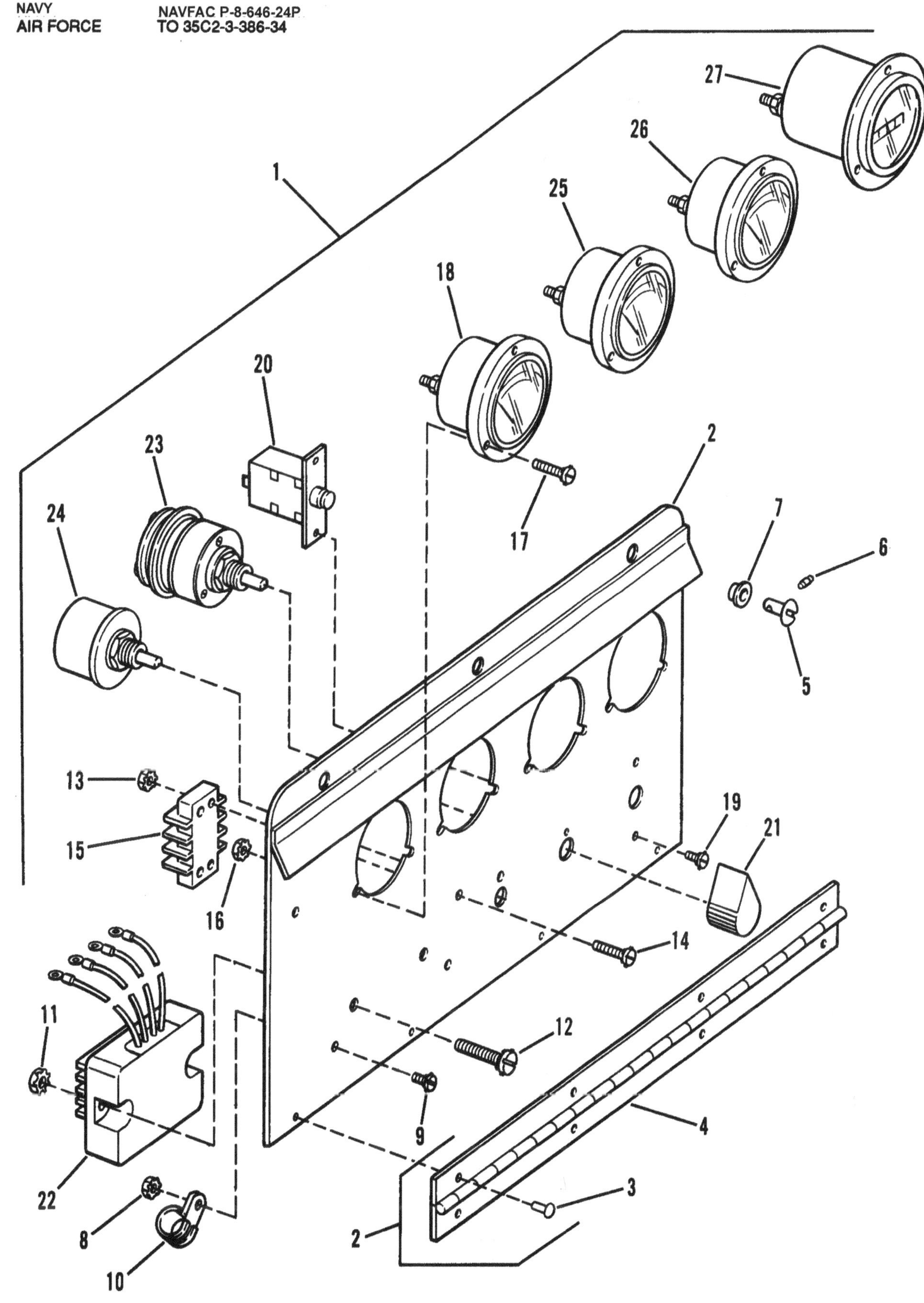

Figure 35. Meter Panel 28 VDC, Group 08 (Generator Controls and Instruments)

(1) ILLUSTRATION (a) FIG NO.	(1) ILLUSTRATION (b)	(2) SMR CODE a ARMY	(2) SMR CODE b AIR FORCE	(2) SMR CODE d NAVY	(2) SMR CODE e USMC	(3) NATIONAL STOCK NUMBER	(4) DESCRIPTION REF NUMBER MFR CODE	USABLE ON CODE	(5) U/M	(6) QTY INC IN	(7) USMC QTY PER EQUIP
							GROUP 08 GENERATOR CONTROLS AND INSTRUMENTS				
35	1		AOOOO				PANEL,CONTROL 301-9323 (44940) 84-13070 (30554)	C	EA 1	1	
35	2		BOFZZ				DOOR,METER 3194-189 (44940) 84-13289 (30554)	C	EA 1	1	
35	3		PAOZZ				RIVET M24243/1-D403 (81349)	C	EA	4	4
35	4		PAOZZ			5320-00-3954523	HINGE MS35823-6C (9696)	C	EA	1	4
35	5		PAOZZ			5340-01-03-7130	STUD 7598948 (19200) MIL-F-5591 (81349)	ABC	EA 3	3	
35	6		PAOZZ			5325-00-432-9899	PIN, GROOVED 69-695 (30554) 1005169 (18876) 99836 (60119)	ABC	EA	3	3
35	7		PAOZZ			5315-00-449-2945	EYELET, METALLIC 295901-2 (60119) 406-0371 (44940) 69-766 (30554)	C	EA 3	3	
35	8		PAOZZ			5325-00-099-8827	NUT, PLAIN, ASSEMBLED 511-10180-00 (78189) 69-561-4 (30554)	C	EA	1	21
35	9		PAOZZ			5310-00094-3421	SCREW, ASSEMBLED P15121-48 (45722) 69-662-48 (30554)	C	EA	1	20
35	10		PAOZZ			530540436-6978	CLAMP,LOOP MS21333-72 (96906)	C	EA 3	1	
35	11		PAOZZ			5340-00491-3790	NUT, PLAIN 501-250800-00 (78189) 69-561-5 (30554)	ABC	EA 2	2	
35	12		PAOZZ			5310696 5173	SCREW, ASSEMBLED P-11t2l-82 (45722) 69-662-82 (30554)	C	EA 4	2	
35	13		PAOZZ			530501-114-5801	NOT, PLAIN 511-O81800-00 (78189) 870-1221 (44940) 69-561-3 (30554)	C	EA	2	50

(1) ILLUS-TRA-TION (a) FIG NO.	(b)	(2) SMR CODE a ARMY	b AIR FORCE	d NAVY	e USMC	(3) NATIONAL STOCK NUMBER	(4) DESCRIPTION REF NUMBER MFR CODE	USABLE ON CODE	(5) U/M	(6) QTY INC IN	(7) USMC QTY PER EQUIP
							GROUP 08 GENERATOR CONTROLS AND INSTRUMENTS				
35	14	PAOZZ	PAOZZ	PAOZZ	PAOZZ	530540G38-3145	SCREW, ASSEMBLED P15121-37 (45722) 69-662-37 (30554)	C	EA	2	5
35	15	PAOZZ	PAOZZ	PAOZZ	PAOZZ	594040D983-6114	BOARD,TERMINAL 38TB3 (81349)	C	EA	1	1
35	16	PAOZZ	PAOZZ	PAOZZ	PAOZZ	531044-3520	NUT, PLAIN, ASSEMBLED 501-040800-00 (78189) 69-561-1 (30554)	C	EA	12	39
35	17	PAOZZ	PAOZZ	PAOZZ	PAOZZ	5305-00-224-1092	SCREW, ASSEMBLED P15121-5 (45722) 69-662-5 (30554)	C	EA	12	24
35	18	PAOZZ	PAOZZ	PAOZZ	PAOZZ		VOLTMETER 13208E8535 (97403)	C	EA	1	1
35	19	PAOZZ	PAOZZ	PAOZZ	PAOZZ	6625-01-157-9516	SCREW, ASSEMBLED P15121-17 (45722) 69-662-17 (30554)	ABC	EA	2	2
35	20	PAOZZ	PAOZZ	PAOZZ	PAOZZ	5305-00-211-9344	BREAKERCIRCUT MS25244-7 1/ 2 (96906)	C	EA	1	3
35	21	PAOZZ	PAOZZ	PAOZZ	PAOZZ	5925-00-686-3290	KNOB MS91528-2K4B (9696)	C	EA	2	10
35	22	PAOZZ	PAOZZ	PAOZZ	PAOZZ	535540-899-9014	REGULATOR, ENGINE 12720 (2N114) 301-0736 (44940) 84-13183 (30554)	C	EA	1	2
35	23	PAOZZ	PAOZZ	PAOZZ	PAOZZ	2920-01-282-8522	SWITCH,ROTARY 72-5011 (30554) 308-0318 (44940) 76902LA (82121)	C	EA	1	2
35	24	PAOZZ	PAOZZ	PAOZZ	PAOZZ	5930-0155-9251	REISTOR,VARIABLE M22-03-00191FD (81349)	ABC	EA	1	3
35	25	PAOZZ	PAOZZ	PAOZZ	PAOZZ		METER, POWER FACTOR 13208E8534 (97403)	C	EA	1	1
35	26	PAOZZ	PAOZZ	PAOZZ	PAOZZ	59054043-5129	INDICATOR, ELECTRICAL 13211E5004 (97403) S1045C (55026)	C	EA	1	1
35	27	PAOZZ	PAOZZ	PAOZZ	PAOZZ	6625-0064-0562	METER, TIME TOTALIZING 15001 (74400) M3971/1-5 (81349)	C	EA	1	2
						6680-00-984-4745					

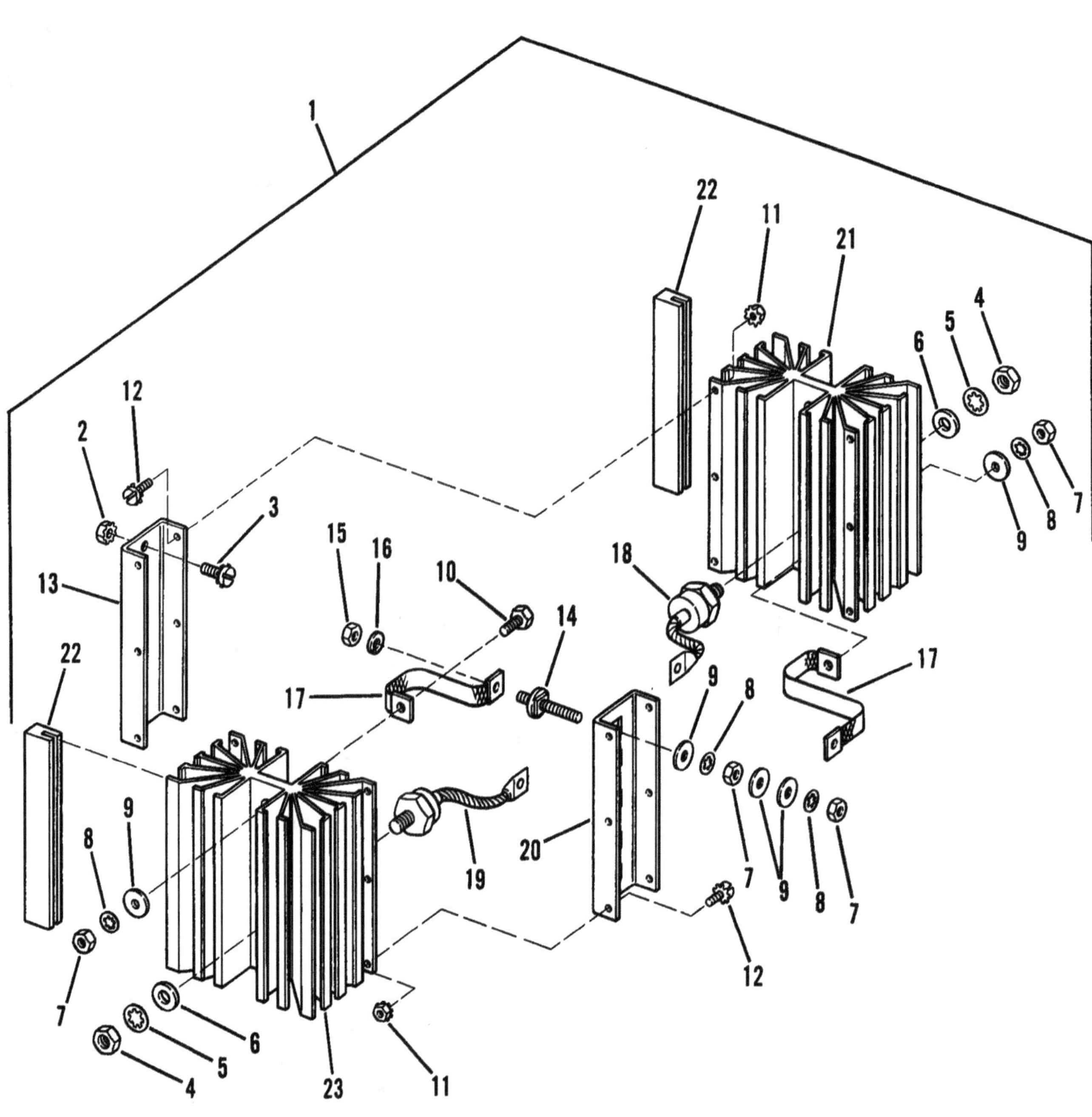

Figure 36.3 Phase Rectifier Bridge 28 VDC, Group 08 (Generator Controls and Instruments)

(1) ILLUSTRATION (a) FIG NO.	(b)	(2) SMR CODE a ARMY	b AIR FORCE	d NAVY	e USMC	(3) NATIONAL STOCK NUMBER	(4) DESCRIPTION REF NUMBER MFR CODE	USABLE ON CODE	(5) U/M	(6) QTY INC IN	(7) USMC QTY PER EQUIP
							GROUP 08 GENERATOR CONTROLS AND INSTRUMENTS				
36	1	PAOOO	PAOOO	PAOOO	PAOOO		RECTIFIER, BRIDGE 305-0733 (44940) 84-13067 (30554)	C	EA 1	1	
36	2	PAOZZ	PAOZZ	PAOZZ	PAOZZ	5310-00-696-5173	NUT, PLAIN, ASSEMBLED 69-561-5 (30554) 501-250800-00 (78189)	C	EA 2	2	
36	3	PAOZZ	PAOZZ	PAOZZ	PAOZZ	5305-00-776-9564	SCREW, ASSEMBLED 42817S8 (23040) ANSI B18.13 (80204)	C	RA 8	2	
36	4	PAOZZ	PAOZZ	PAOZZ	PAOZZ	531004732-M59	NUT, PLAIN, REXAM MS19684 (96906)	C			
36	5	PAOZZ	PAOZZ	PAOZZ	PAOZZ		WASHER, LOCK MS35333-42 (96906)	C	EA 6	6	
36	6	PAOZZ	PAOZZ	PAOZZ	PAOZZ	5310-595-7237	WASHER,FLAT ANW60616L (811044)	C	EA 6	6	
36	7	PAOZZ	PAOZZ	PAOZZ	PAOZZ	5310-0167-837	NUT,HEX PLAIN MS35649-2255N (96906)	- C	EA 6	6	
36	8	PAOZZ	PAOZZ	PAOZZ	PAOZZ	5310-00-732-0559	WASHER,LOCK MS35333-108 (96906)	C			
36	9	PAOZZ	PAOZZ	PAOZZ	PAOZZ	531-00-022-8834	WASHER,FLAT(BRASS) MS15795-910 (96906)	C	EA 12	37	
36	10	PAOZZ	PAOZZ	PAOZZ	PAOZZ	5310-00-045-5210	SCREW.CAP HH MS35309-306 (96906)	C	EA 12	12	
36	11	PAOZZ	PAOZZ	PAOZZ	PAOZZ	2510-00-534-5828	NUT, PLAIN, ASSEBLED 511-081800-00 (78189) 870-1221 (44940) 69-561-3 (30554)	C	EA 17	17	
									EA 2	2	
36	12	PAOZZ	PAOZZ	PAOZZ	PAOZZ	5310-00-052-3632	SCREW, ASSEMBLED PI5121-34 (45722) 69-662-34 (30554)	C	EA 12	12	
36	13	PAOZZ	PAOZZ	PAOZZ	PAOZZ	5305-00-038-3122	BRACKET,RECTIFIER 305-0771 (44940) 84-13054 (30554)	C	EA 12	12	
36	14	PAOZZ	PAOZZ	PAOZZ	PAOZZ		TERMINAL, RECTIFIER 305-0774 (44940) 84-13057 (30554)	C	EA 1	1	
36	15	PAOZZ	PAOZZ	PAOZZ	PAOZZ		NUT, PLAIN,HEXAGON MS51967-2 (96906) 862-1001 (44940)	C	EA	5	5
									EA 5	5	

(1) ILLUSTRATION (a) FIG NO.	(b)	(2) SMR CODE a ARMY	b AIR FORCE	d NAVY	e USMC	(3) NATIONAL STOCK NUMBER	(4) DESCRIPTION REF NUMBER MFR CODE	USABLE ON CODE	(5) U/M	(6) QTY INC IN	(7) USMC QTY PER EQUIP
							GROUP 08 GENERATOR CONTROLS AND INSTRUMENTS				
36	16	PAOZZ	PAOZZ	PAOZZ	PAOZZ	5310-00-187-2425	WASHER,LOCK MS35338-120 (96906)	C	EA	10	22
36	17	PAOZZ	PAOZZ	PAOZZ	PAOZZ		DC CONNECR RECT................................ 305n3 (44940) 84-13056 (30554)	C	EA	2	2
36	18	PAOZZ	PAOZZ	PAOZZ	PAOZZ		RECTIFIER.. S6874N-2 (51589) 84-13162 (30554)	C	EA	3	3
36	19	PAOZZ	PAOZZ	PAOZZ	PAOZZ		RECTIFIER.. S6874P-2 (5159) 84-13068-1 00554)	C	EA	3	3
36	20	PAOZZ	PAOZZ	PAOZZ	PAOZZ		TERMINAL SUPPORT,RE......................... 305-472 (44940) 4-13055 (30554)	C	EA	1	1
36	21	XBOZZ	XBOZZ	XBOZZ	XBOZZ		HEAT SINK, NEG 305-o76 (44940) 84-13059 (30554)	C	EA	1	1
36	22	PAOZZ	PAOZZ	PAOZZ	PAOZZ		DAMPER,VIBRATION 3054775 (44940) 84-1305S (30554)	C	EA	2	2
36	23	XBOZZ	XBOZZ	XBOZZ	XBOZZ		HEAT SINK,POSITIVE............................... 363-0098 (44940) 84-13187 (30554)	C	EA	1	1

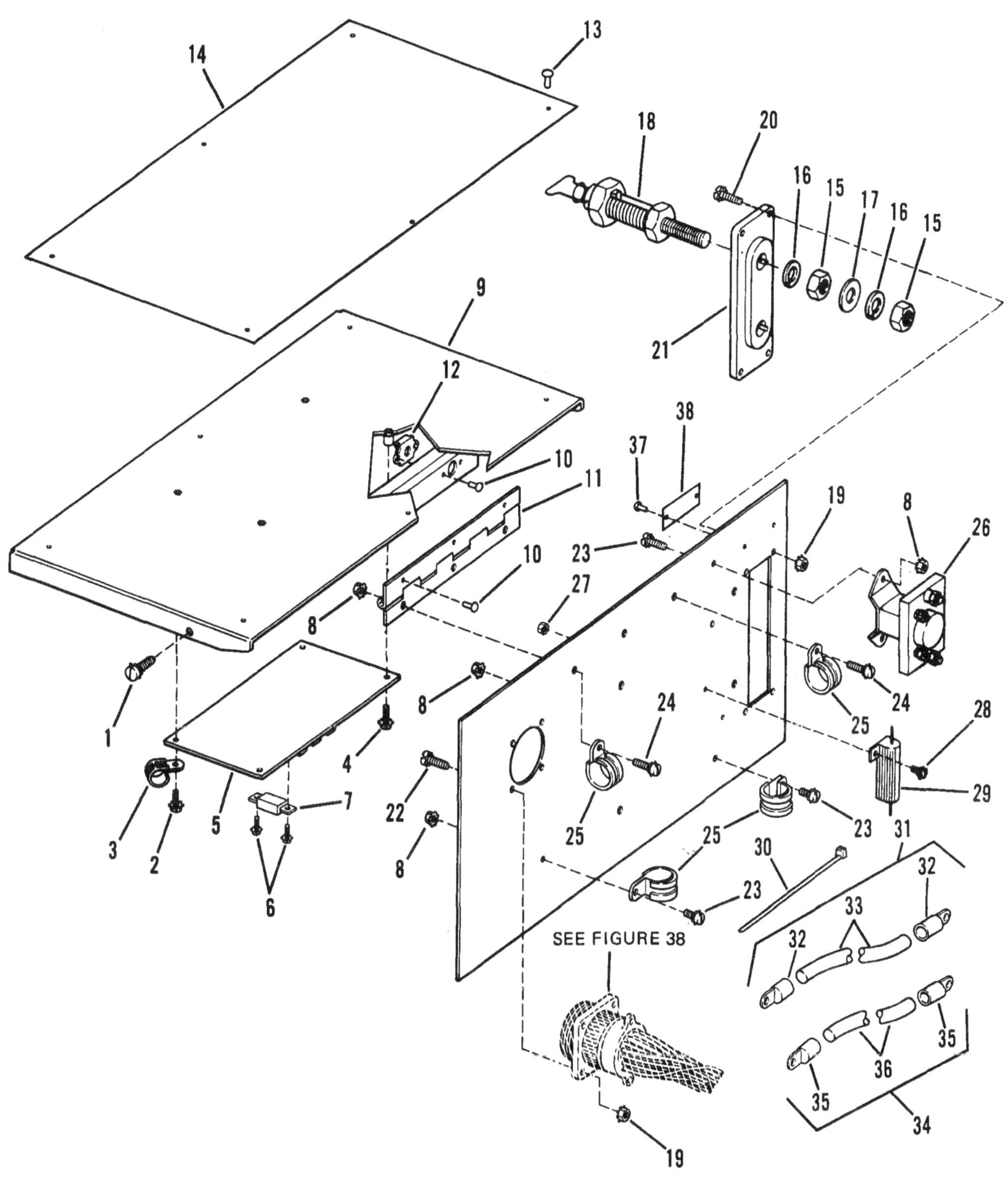

Figure 37. Generator Control Assembly 28 VDC, Group 08 (Generator Controls and Instruments) (Sheet 1 of 3)

(1) ILLUSTRATION (a) FIG NO.	(b)	(2) SMR CODE a ARMY	b AIR FORCE	d NAVY	e USMC	(3) NATIONAL STOCK NUMBER	(4) DESCRIPTION REF NUMBER MFR CODE / USABLE ON CODE	(5) U/M	(6) QTY INC IN	(7) USMC QTY PER EQUIP
							Group 08 GENERATOR CONTROLS AND INSTRUMENTS			
37	1	PAOZZ	PAOZZ	PAOZZ	PAOZZ	5305 00-776-9564	SCREW, ASSEMBLED .... C 4217S8 (23040) ANSI B18.13 (80204)	EA 11 11		
37	2	PAOZZ	PAOZZ	PAOZZ	PAOZZ		SCREW, ASSEMBLED .... C P15121-20 (45722) 69-62-20 (30554)	EA	1	9
37	3	PAOZZ	PAOZZ	PAOZZ	PAOZZ		CLAMP, LOOP .... C MS21333-72 (96906)	EA 2	2	
37	4	PAOZZ	PAOZZ	PAOZZ	PAOZZ		SCREW, ASSEMBLED .... C P15121-21 (45722) 69-662-21 (30554)	EA	5	10
37	5	PAFZZ	**PAFZZ**	PAFZZ	PAFZZ		CIRCUIT CARD ASSY.......... C 300-2953 (44940) 84-13178 (30554)	EA 2	1	
37	6	PAFZZ	**PAFZZ**	PAFZZ	**PAFZZ**		SCREW, ASSEMBLED .... C P15121-2 (45722) 69-662-2 (30554)	EA	8	10
37	7	PAFZZ	PAFZZ	PAFZZ	PAFZZ		RELAY.......... C M5757/23-003 (81349)	EA 8	4	
37	8	PAOZZ	PAOZZ	PAOZZ	PAOZZ		NUT, PLAIN, ASSEMBLED .... C 511-101800-4 (78189) 69-561-4 (30554)	EA 18 18		
37	9	XBOZZ	*XBOZZ*	XBOZZ	XBOZZ		COVER, CONTROL BOX .... C 319-190 (44940) 84-13292 (30554)	EA 1	1	
37	10	PAFZZ	PAOZZ	PAOZZ	PAOZZ		RIVET.......... C MS20426B4-6 (96906)	EA 14 14		
37	11	PAFZZ	PAOZZ	PAOZZ	PAOZZ		HINGE, BUTT.......... C MS35823-6C (96906)	EA 1	1	
37	12	PAFZZ	PAOZZ	PAOZZ	PAOZZ		RECEPTACLE, TURNLOCK .... C 99947P130 (61864)	EA 6	3	
37	13	PAFZZ	PAOZZ	PAOZZ	PAOZZ		RIVET.......... C M24243/1-D403 (81349)	EA 6	6	
37	14	MDFZZ	MDOZZ	MDOZZ	MDOZZ		PLATE, SCHEMATIC .... C 099-2335 (44940) 84-13146 (30554)	EA 1	1	

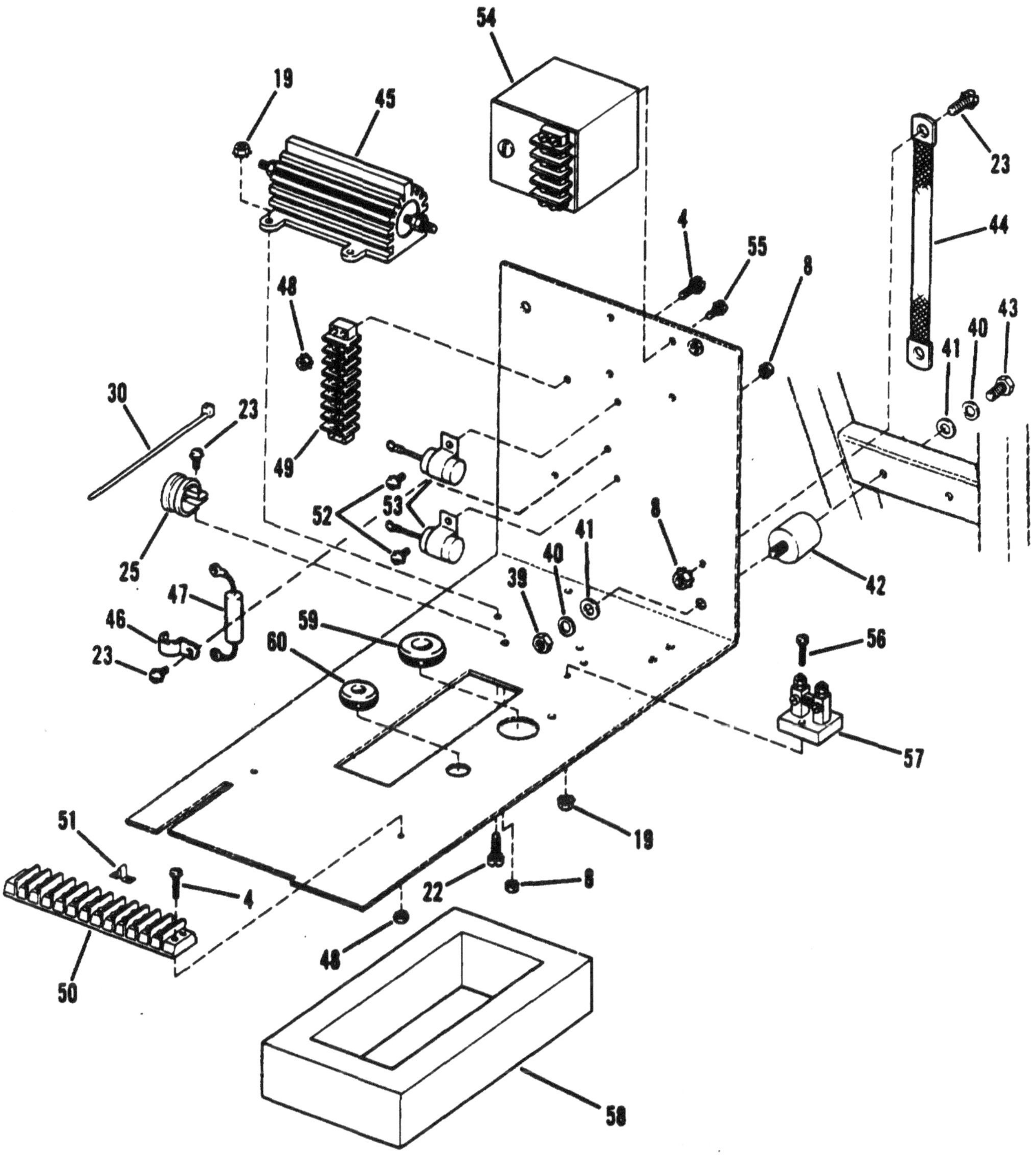

Figure 37. Generator Control Assembly 28 VDC, Group 08 (Generator Controls and Instruments) (Sheet 2 of 3)

(1) ILLUSTRATION (a) FIG NO.	(1) ILLUSTRATION (b)	(2) SMR CODE a ARMY	(2) SMR CODE b AIR FORCE	(2) SMR CODE d NAVY	(2) SMR CODE e USMC	(3) NATIONAL STOCK NUMBER	(4) DESCRIPTION / REF NUMBER MFR CODE	(4) USABLE ON CODE	(5) U/M	(6) QTY INC IN	(7) USMC QTY PER EQUIP
							GROUP 08 GENERATOR CONTROLS AND INSTRUMENTS				
37	15	PAOZZ	PAOZZ	PAOZZ	PAOZZ	5310-00-897-6082	NUT, PLAIN, EXAGON MS35691-36 (96906)	C	EA	4	4
37	16	PAOZZ	PAOZZ	PAOZZ	PAOZZ	5310-00-187-2429	WASHER, LOCK MS35338-124 (96906)	C	EA	4	4
37	17	PAOZZ	PAOZZ	PAOZZ	PAOZZ	5310-01-212-4612	WASHER, FLAT MS15795-417 (96906)	C	EA	2	2
37	18	PAOZZ	PAOZZ	PAOZZ	PAOZZ	5940-00-958-1214	TERMINAL, STUD JHP112-53 (02032) 13208E5820-1 (97403)	C	EA	2	2
37	19	PAOZZ	PAOZZ	PAOZZ	PAOZZ	5310-00-052-3632	NUT, PLAIN, ASSEMBLED 511-081800-00 (78189) 870-1221 (44940) 69-561-3 (30554)	C	EA	16	16
37	20	PAOZZ	PAOZZ	PAOZZ	PAOZZ	5305-00-958-5477	SCREW MS35190-254 (96906)	C	EA	4	8
37	21	PAOZZ	PAOZZ	PAOZZ	PAOZZ	5970-00-929-5627	INSULATOR, PLATE 13216E3987 (97403)	C	EA	1	1
37	22	PAOZZ	PAOZZ	PAOZZ	PAOZZ	5305-00-038-3103	SCREW, ASSEMBLED P15121-35 (45722) 69-662-35 (30554)	C	EA	7	14
37	23	PAOZZ	PAOZZ	PAOZZ	PAOZZ	5305-00-036-6978	SCREW, ASSEMBLED P15121-48 (45722) 69-662-48 (30554)	C	EA	12	12
37	24	PAOZZ	PAOZZ	PAOZZ	PAOZZ	5305-00-036-6906	SCREW, ASSEMBLED P-15121-50 (45722) 69-662-50 (30554)	ABC	EA	2	2
37	25	PAOZZ	PAOZZ	PAOZZ	PAOZZ	5340-00-050-2740	CLAMP.LOOP MS21333-75 (96906)	C	EA	6	8
37	26	PAOZZ	PAOZZ	PAOZZ	PAOZZ	5945-00-686-6877	RELAY MS24166-DI (96906)	ABC	EA	1	2
37	27	PAOZZ	PAOZZ	PAOZZ	PAOZZ	5310-00-836-3520	NUT, PLAIN, ASSEMBLED 501-040800-00 (78189) 69-561-1 (30554)	C	EA	1	1
37	28	PAOZZ	PAOZZ	PAOZZ	PAOZZ	5305-00-036-6970	SCREW, ASSEMBLED P15121-2 (45722) 69-662-2 (30554)	C	EA	2	2

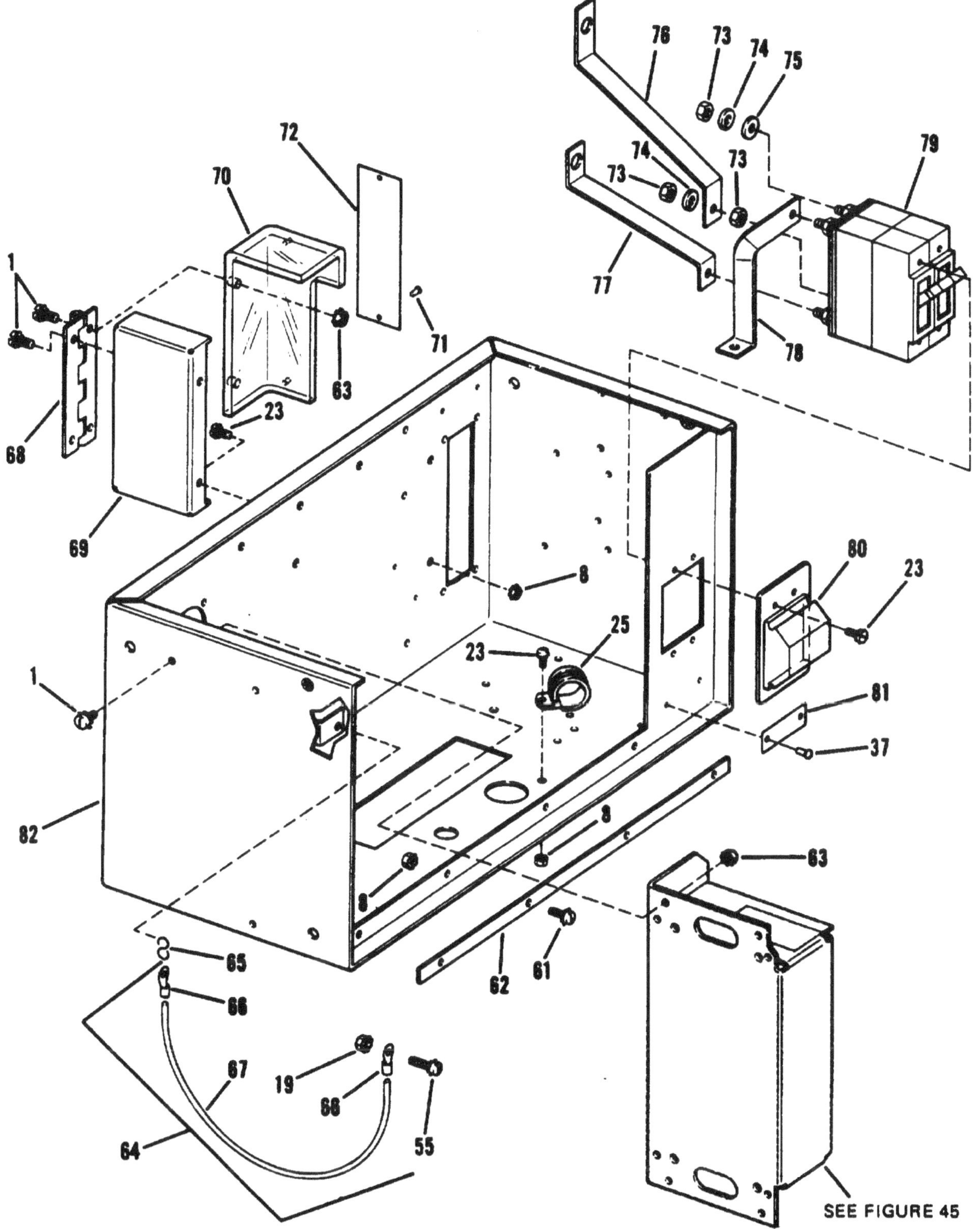

Figure 37. Generator Control Assembly 28 VDC, Group 08 (Generator Controls and Instruments) (Sheet 3 of 3)

(1) ILLUSTRATION (a) FIG NO.	(b)	(2) SMR CODE a ARMY	b AIR FORCE	d NAVY	e USMC	(3) NATIONAL STOCK NUMBER	(4) DESCRIPTION REF NUMBER MFR CODE / USABLE ON CODE	(5) U/M	(6) QTY INC IN	(7) USMC QTY PER EQUIP
							GROUP 08 GENERATOR CONTROLS AND INSTRUMENTS			
37	29	PAOZZ	PAOZZ	PAOZZ	PAOZZ	5950-139-1989	RESISTOR, FIXED, WIRE.... C RER75F4R02R (81349)	EA 2	1	
37	30	PAOZZ	PAOZZ	PAOZZ	PAOZZ	5975-00-074-2072	STRAP, TIE DOWN.... C MS3367-1-9 96906)	EA 30 70		
37	31	AFOOO	AFOOO	AFOOO	AFOOO		LEAD, ASSY.... C 226-3695 (44940) 84-13303 (30554)	EA 1	1	
37	32	PAOZZ	PAOZZ	PAOZZ	PAOZZ	5940-00-115-5008	LUG.TERMINAL .... C MS25036-129 (96906)	EA 4	2	
37	33	PAOZZ	PAOZZ	PAOZZ	PAOZZ	6145-00--578-6597	WIRE .... C M5086/2-1-9 (81349)			
37	34	AFOOO	AFOOO	AFOOO	AFOOO		LEAD ASSY.... C 226-3693 (44940) 84-13301 (30554)	FT	1	2
37	35	PAOZZ	PAOZZ	PAOZZ	PAOZZ		LUG, TERMINAL .... C MS25036-129 (96906)	EA 1	1	
37	36	PAOZZ	PAOZZ	PAOZZ	PAOZZ	5940-00-115-5008	WIRE, ELECT.... C M55086/2-1-9 (81349)	EA 2	2	
37	37	PAOZZ	PAOZZ	PAOZZ	PAOZZ	6145-00-578-6597	RIVET.... ABC RV630-4-2 (53551) 818-0076 (44940) 72-5018 (30554)	FT 1	1	
37	38		MDOZZ			5320-01-049-8263	PLATE, IDENT ....AB 13211E6888-2 (97403)	EA 4	4	
37	39	MDOZZ	PAOZZ	MDOZZ	MDOZZ		NUT, PLAIN, HEXAGON .... ABC MS51967-6 (96906)			
37	40	PAOZZ	PAOZZ	PAOZZ	PAOZZ		WASHER, LOCK .... ABC MS35338-45 (96906) 850-1045 (44940)	EA	1	2
37	41	PAOZZ	PAOZZ	PAOZZ	PAOZZ	5310-00-931-8167	WASHER, FLAT MS27183-12 (96906)	EA 4	4	
37	42	PAOZZ	PAOZZ	PAOZZ	PAOZZ	5310-00407-9566	MOUNT, RESILIENT .... ABC 29031 (81860) 84-13015 (30554)	EA 8	8	
37	43	PAOZZ	PAOZZ	PAOZZ	PAOZZ	531M-0014219	BOLT, MACHINE.... ABC MS90725-31 (96906)	ABC EA a8		
		PAOZZ		PAOZZ	PAOZZ	5340-01-275-3534		EA 8	4	
						5306-00-225-1496		EA 4	4	

(1) ILLUSTRATION (a) FIG NO.	(b) ITEM NO.	(2) SMR CODE a ARMY	b AIR FORCE	d NAVY	e USMC	(3) NATIONAL STOCK NUMBER	(4) DESCRIPTION REF NUMBER MFR CODE	USABLE ON CODE	(5) U/M	(6) QTY INC IN	(7) USMC QTY PER EQUIP
							GROUP 08 GENERATOR CONTROLS AND INSTRUMENTS				
37	44	PAOZZ	PAOZZ	PAOZZ	PAOZZ	5340-01-233-8305	STRAP, GROUND 72-5006-1 (30554)	C	EA	1	2
37	45	PAOZZ	PAOZZ	PAOZZ	PAOZZ	5905-00-883-8431	RESISTOR, FIXED, WIRE RE80G1R00 (31349)	C	EA	1	2
37	46	PAOZZ	PAOZZ	PAOZZ	PAOZZ	5340-00-764-7051	CLAMP, LOOP MS21333-69 (96906)	C	EA	1	1
37	47	PAOZZ	PAOZZ	PAOZZ	PAOZZ		CAPACITOR 356-0121 (44940) 84-13086 (30554)	C	EA	1	1
37	48	PAOZZ	PAOZZ	PAOZZ	PAOZZ	5310-00-063-7360	NUT, PLAIN, ASSEMBLED 511-061800-00 (78189) 69-561-2 (30554)	C	EA	4	20
37	49	PAOZZ	PAOZZ	PAOZZ	PAOZZ	5940-00-983-6049	BOARD, TERMINAL 37TB8 (81349)	C	EA	1	1
37	50	PAOZZ	PAOZZ	PAOZZ	PAOZZ	5940-00-983-6089	BOARD, TERMNAL 38TB12 (81349)	C	EA	1	1
37	51	PAOZZ	PAOZZ	PAOZZ	PAOZZ	6150-00-519-2714	BUS, CONDUCTOR M55164/28A-TBJB (81349)	C	EA	3	6
37	52	PAOZZ	PAOZZ	PAOZZ	PAOZZ	5305-00-036-6968	SCREW, ASSEMBLED WASHER P15121-18 (45722) 69-662-18 (30554)	C	2	2	
37	53	PAOZZ	PAOZZ	PAOZZ	PAOZZ		CAPACITOR 312-0252 (44940) 84-13071 (3554)	C	EA	2	2
37	54	PAOZZ	PAOZZ	PAOZZ	PAOZZ	6125-00-659-2935	FREQUENCY CONVERTER 11370 (63638) 13213E4237 (97403)	C	EA	1	1
37	55	PAOZZ	PAOZZ	PAOZZ	PAOZZ	5305-00-036-6977	SCREW, ASSEMBLED P15121-33 (45722) 69-662-33 (30554)	C	EA	5	16
37	56	PAOZZ	PAOZZ	PAOZZ	PAOZZ	5305-00-0-038-3145	SCREW, ASSEMBLED P15121-37 (45722) 69-662-37 (30554)	C	EA	1	1

(1) ILLUSTRATION (a) FIG NO.	(b)	(2) SMR CODE a ARMY	b AIR FORCE	d NAVY	e USMC	(3) NATIONAL STOCK NUMBER	(4) DESCRIPTION REF NUMBER MFR CODE / USABLE ON CODE	(5) U/M	(6) QTY INC IN	(7) USMC QTY PER EQUIP
							GROUP 08 GENERATOR CONTROLS AND INSIRUMENTS			
37	57	PAOZZ	PAOZZ	PAOZZ	PAOZ	6625-00-055-9760	SHUNT, INSTRUMENT.......... C 13213E4264 (97403)	EA	1	1
37	58	XBOZZ	XBOZZ	XBOZZ	XBOZZ		DUCT-AIR, SPONGE .... C 134-4543 (44940) 84-13181 (30554)	EA 2	1	
37	59	PAOZZ	PAOZZ	PAOZZ	PAOZZ	5325-00-754-2187	GROMMET.......... C MS35489-123 (96906)	EA 2	1	
37	60	PAOZZ	PAOZZ	PAOZZ	PAOZZ	5325-00-185-0003	GROMMET.......... C 2860 (70485) 205K1114' (94990)	EA 1	1	
37	61	PAOZZ	PAOZZ	PAOZZ	PAOZZ		SCREW, ASSEMBLED WASHER .... C 303737 (32195)			
						5305-00-177-5778		EA 9	4	
37	62	XBOZZ	XBOZZ	XBOZZ	XBOZZ		PLATE, SPACER.......... C 301-9471 (44940) 84-13144 (30554)			
37	63	PAOZZ	PAOZZ	PAOZZ	PAOZZ		NUT, PLAIN, ASSEMBLED .... ABC 501-250800-00 (78189) 69-561-5 (30554)	EA	1	1
						5310-00-696-5173		EA 7	7	
37	64	AFOZZ	AFOZZ	AFOZZ	AFOZZ		HOLDER, METER PANEL.......... C 406-0583 (44940) 84-13069 (30554)			
37	65	PAOZZ	PAOZZ	PAOZZ	PAOZZ		HOOK.CHAINS.......... ABC MS87006-13 (969016)	EA 1	1	
37	66	PAOZZ	PAOZZ	PAOZZ	PAOZZ	4030-00-780-9350	TERMINAL, LUG, CRIMP.......... ABC MS20659-144 (96906)			
								EA 2	1	
37	67	PAOZZ	PAOZZ	PAOZZ	PAOZZ		ROPE, NYLON .... ABC 8198810 (19200)			
						5940-00-115-2677				
37	68	PAOZZ	PAOZZ	PAOZZ	PAOZZ		HINGE, BUTT.......... C 406-0581 (44940) 84-13190 (30554)	EA 4	2	
						4020-00-523-9641				
37	69	XBOZZ	XBOZZ	XBOZZ	XBOZZ		BRACKET, MOUNTING .... C 319-0194 (44940) 84-13293 (30554)	IN	9 18	
						5340-01-281-5270		EA 2	1	
37	70	PAOZZ	PAOZZ	PAOZZ	PAOZZ		COVER, ELECTRICAL .... C 332-2767 (44940) 84-13171 (30554)			
								EA 2	1	
						5935-01-280-1177		EA	1	2

(1) ILLUSTRATION (a) FIG NO.	(b) NO.	(2) SMR CODE a ARMY	b AIR FORCE	d NAVY	e USMC	(3) NATIONAL STOCK NUMBER	(4) DESCRIPTION REF NUMBER MFR CODE	USABLE ON CODE	(5) U/M	(6) QTY INC IN	(7) USMC QTY PER EQUIP
							GROUP 08 GENERATOR CONTROLS AND INSTRUMENTS				
37	71	PAFZZ	PAOZZ	PAOZZ	PAOZZ	532040.971-771	RIVET M24243/1-D405 (81349)	C	EA		4
37	72	MDFZZ	MDOZZ	MDOZZ	MDOZZ		PLATE, TERM. ID 099-2382 (44940) 84-13296 (30554)	C	EA		1
37	73	PAOZZ	PAOZZ	PAOZZ	PAOZZ	5310-01-301-1982	NUT, HEX MS35649-2255N (96906)	C	EA		16
37	74	PAOZZ	PAOZZ	PAOZZ	PAOZZ	5310-00-187-2425	WASHER, LOCK MS35338-120 (96906)	C	EA		12
37	75	PAOZZ	PAOZZ	PAOZZ	PAOZZ	5310-00-950-0440	WASHER, FLAT MS15795-409 (96906)	C	EA		6
37	76	XBOZZ	XBOZZ	XBOZZ	XBOZZ		BUS BAR 337-2248 (44940) 84-13196 (30554)	C	EA		1
37	77	XBOZZ	XBOZZ	XBOZZ	XBOZZ		BUS BAR 337-2248 (44940) 84-13195 (30554)	C	EA		1
37	78	XBOZZ	XBOZZ	XBOZZ	XBOZZ		BUS BAR 337-2249 (44940) 84-13197 (30554)	C	EA		1
37	79	PAOZZ	PAOZZ	PAOZZ	PAOZZ	5925-00-961-1202	BREAKER, CIRCUIT 13208E5865 (97403) 52-132-1MG3 (74193)	C	EA		1
37	80	PAOZZ	PAOZZ	PAOZZ	PAOZZ		BOOT, DUST AND MOISTURE 006-10196 (74193)	C	EA		1
37	81	MDFZZ	MDOZZ	MDOZZ	MDOZZ	5975-00-136-9005	PLATE,.ID 13211E66888-1 (97403)	C	EA		1
37	82	XBOZZ	XBOZZ	XBOZZ	XBOZZ	9905-00-477-4131	BOX, CONTROL 301-9273 (44940) 84-13176 (30554)	C	EA		1

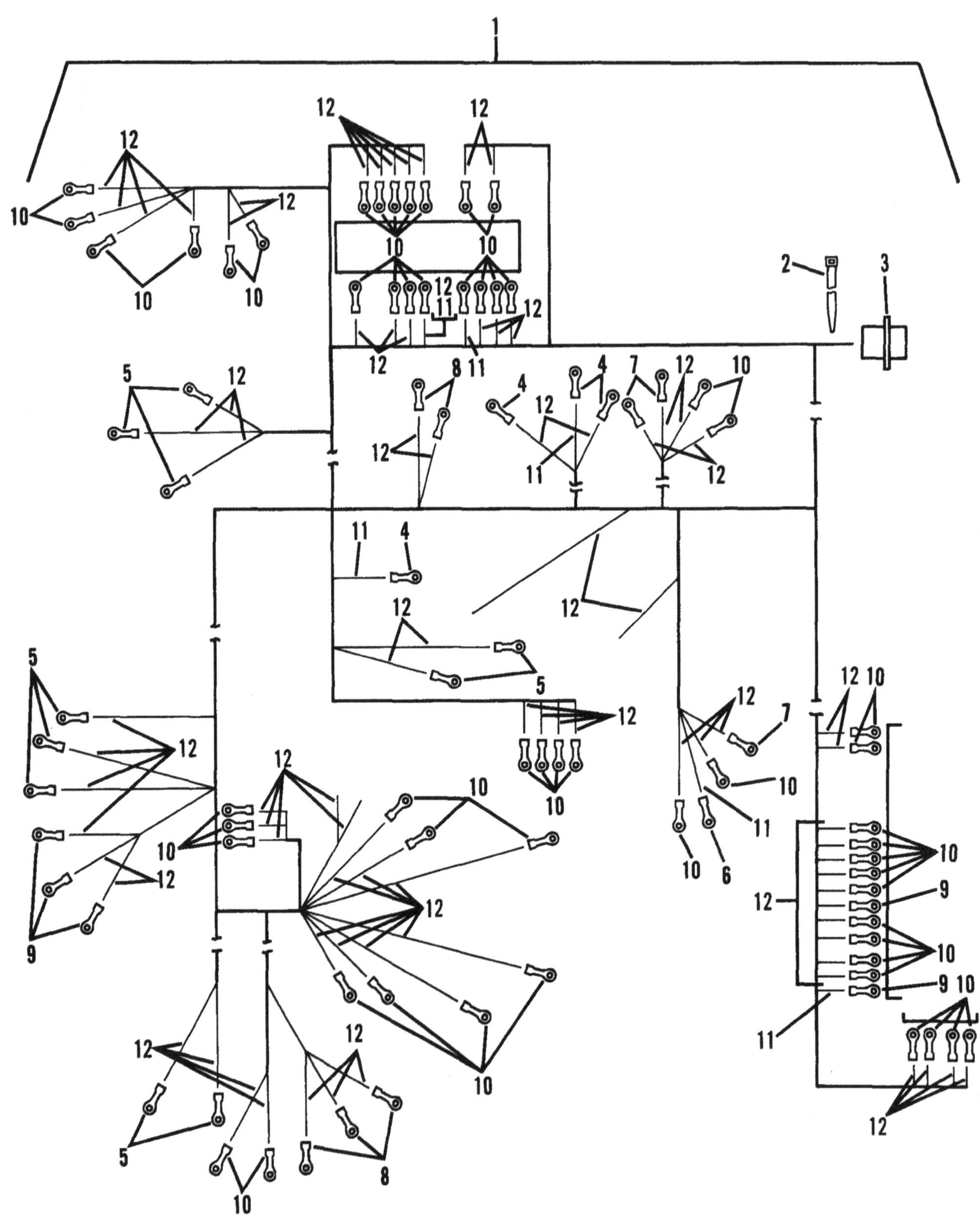

Figure 38. Control Box Wire Harness 28 VDC, Group 08 (Generator Controls and Instruments)

(1) ILLUSTRATION (a) FIG NO.	(1) ILLUSTRATION (b) ITEM NO.	(2) SMR CODE a ARMY	(2) SMR CODE b AIR FORCE	(2) SMR CODE d NAVY	(2) SMR CODE e USMC	(3) NATIONAL STOCK NUMBER	(4) DESCRIPTION REF NUMBER MFR CODE	(4) USABLE ON CODE	(5) U/M	(6) QTY INC IN	(7) USMC QTY PER EQUIP
							GROUP 08 GENERATOR CONTROLS AND INSTRUMENTS				
38	1		AOOOO				HARNESS, WIRING 338-1899 (44940) 84-13156 (30554)	C	EA	1	1
38	2		PAOZZ			5975-00-111-3208	STRAP, TIEDOWN MS3367-5-9 (96906)	C	EA	80	160
38	3		PAOZZ			5935-01-219-4205	CONNECTOR, RECEPT MS3450W32-7P (96906)	C	EA	1	2
38	4		PAOZZ			5940-00-143-4777	TERMINAL, LUG MS25036-157 (96906)	C	EA	3	22
38	5		PAOZZ			5940-00-230-0515	TERMINAL, LUG MS25036-154 (96906)	C	EA	17	32
38	6		PAOZZ			5940-00-1434794	TERMINAL, LUG MS25036-112 (96906)	C	EA	1	1
38	7		PAOZZ			5940-00-143-4780	TERMINAL, LUG MS25036-108 (96906)	C	EA	2	3
38	8		PAOZZ			59400-143-4774	TERMINAL, LUG MS25036-153 (96906)	C	EA	5	18
38	9		PAOZZ			5940-00-204-8990	TERMINAL, LUG MS25036-111 (96906)	C	EA	2	2
38	10		PAOZZ			594040-283-5280	TERMINAL, LUG MS25036-106 (96906)	C	EA	4	105
38	11		PAOZZ			6145-00-578-7514	WIRE, ELECTRICAL M5086/6-12-9 (81349)	C	FT	16	16
38	12		PAOZZ			6145-00-655-2562	WIRE, ELECTRICAL M5086/3-16-9 (81349)	C	FT	16	16

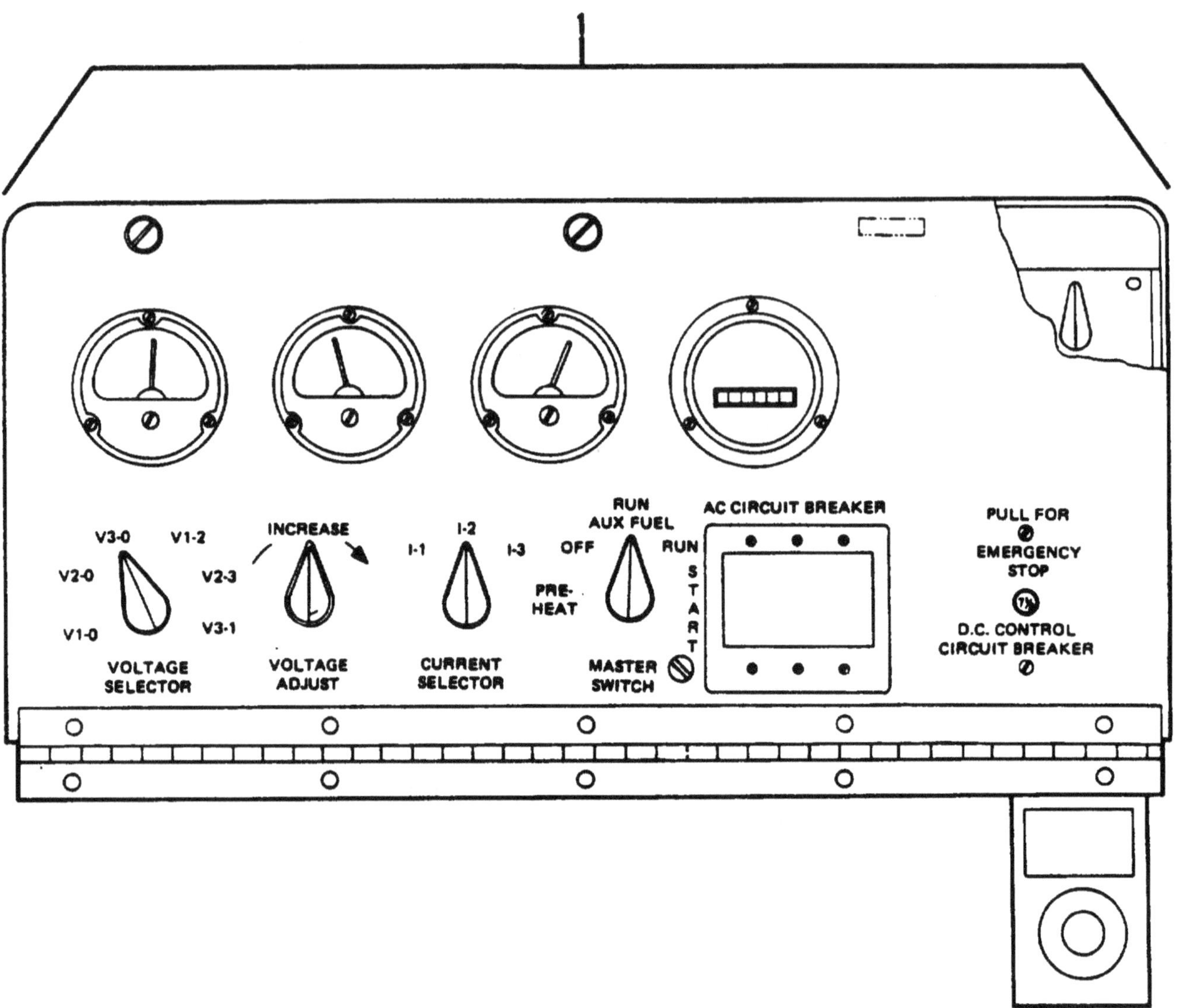

Figure 39. Generator Control 60 Hz, Group 08 (Generator Controls and Instruments)

(1) ILLUSTRATION FIG NO	(2) ITEM NO	(3) SMR CODE ARMY	AIR FORCE	NAVY	USMC	(4) NATIONAL STOCK NUMBER	(5)(6)(7) DESCRIPTIION REF NUMBER MFR CODE	USABLE ON CODE	U/M	INC IN UNITS	QTY PER EQUIP
							GROUP 08 GENERATOR CONTROLS AND INSTRUMENTS				
39	1	XBOOO	XBOOO	XBOOO	XBOOO		CONTROL,60 300-2963-01 (44940) 84- 13089-01 (30554)	A	EA	1	1

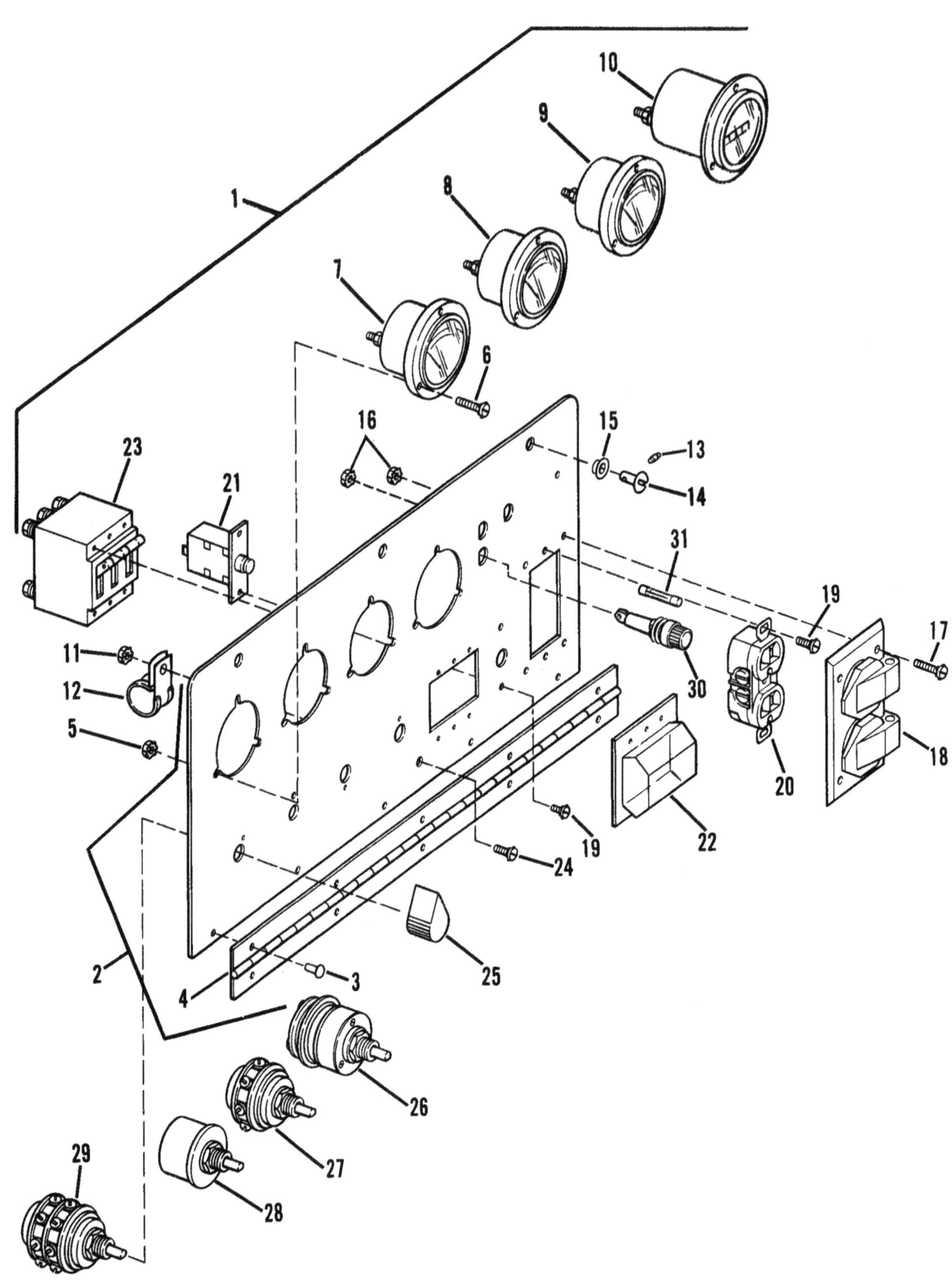

Figure 40. Meter Panel 60 Hz, Group 08 (Generator Controls and Instruments)

(1) ILLUSTRATION (a) FIG NO.	(1) ILLUSTRATION (b)	(2) SMR CODE a ARMY	(2) SMR CODE b AIR FORCE	(2) SMR CODE d NAVY	(2) SMR CODE e USMC	(3) NATIONAL STOCK NUMBER	(4) DESCRIPTION / REF NUMBER MFR CODE	USABLE ON CODE	(5) U/M	(6) QTY INC IN	(7) USMC QTY PER EQUIP
							GROUP 08 GENERATOR CONTROLS AND INSTRUMENTS				
40	1		AOOOO		AOOOO		PANEL, CONTROL 301-9324 (44940) 84-13088 (30554)	A	EA	1	1
40	2		XBOZZ		XBOZZ		DOOR, METER 319-0186 (44940) 84-13281 (30554)	A	EA	1	1
40	3		PAOZZ		PAOZZ	5320-00-395-6523	RIVET M24243/1-D403 (81349)	A	EA	5	5
40	4		PAOZZ		PAOZZ	5340-01-053-7130	HINGE MS35823-6C (96906)	A	EA	1	1
40	5		PAOZZ		PAOZZ	531000836-3520	NUT, PLAIN 50m100Mo(o4 (78139) 69-561-1 (30554)	ABC	EA	12	12
40	6		PAOZZ		PAOZZ	530540-224-1092	SCREW, ASSEMBLED P15121-5 (45722) 69662-5 (30554)	ABC	EA	12	12
40	7		PAOZZ		PAOZZ	6625-00-065-5301	VOLTMETER 13211E6905 (97403)	AA	EA	1	1
40	8		PAOZZ		PAOZZ	6625-054-2038	METER.FREQUENCY 13211E6992-1 (97403) S-10448 (55026)	A	EA	1	1
40	9		PAOZZ		PAOZZ	66254065-5258	METER, ARBITRARY 1321 IE6919 (97403)	AB	EA	1	1
40	10		PAOZZ		PAOZZ	6645400-9-8342	METER TIMETOTALIZI M397111-5 (81349) 15001 (74400)	ABC	EA	1	1
40	11		PAOZZ		PAOZZ		NUT, & CAP WASHER 511-10100e4 (78139) S70-1232 (44940) 69-561-5 (30554)	ABC	EA	1	1
40	12		PAOZZ		PAOZZ	5340-00-702-2z48	CLAMP, LOOP MS21333-12S (96906)	ABC	FA	1	3
40	13		PAOZZ		PAOZZ	5315-00-449-2945	PIN, GROOVED 99836 (60119) 69-695 (30554) 1005169 (18876)	ABC	EA	3	3

(1) ILLUSTRATION (a) FIG NO.	(b)	(2) SMR CODE a ARMY	b AIR FORCE	d NAVY	e USMC	(3) NATIONAL STOCK NUMBER	(4) DESCRIPTION / REF NUMBER MFR CODE	USABLE ON CODE	(5) U/M	(6) QTY INC IN	(7) USMC QTY PER EQUIP
							GROUP 08 GENERATOR CONTROLS AND INSTRUMENTS				
40	14	PAOZZ	PAOZZ	PAOZZ	PAOZZ	5325-00-432-9899	STUD 92-92-2-22-0 (60119) MIL-F-5591 (81349:	ABC	EA	3	3
40	15	PAOZZ	PAOZZ	PAOZZ	PAOZZ	5325-00-099-8827	EYELET, METALLIC 69-766 (30554) 406-0371 (44940) 295901-2 (60119)	ABC	EA	3	3
40	16	PAOZZ	PAOZZ	PAOZZ	PAOZZ	5310-00-063-7360	NUT, PLAIN 511-061800-00 (78189) 69-561-2 (30554)	ABC	EA	6	6
40	17	PAOZZ	PAOZZ	PAOZZ	PAOZZ	5305-00-036-6972	SCREW, ASSEMBLED P15121-20 (45722) 69-662-20 (30554)	ABC	EA	4	4
40	18	PAOZZ	PAOZZ	PAOZZ	PAOZZ	5975-00-879-7234	PLATE, WALL, ELEC 5211 (74545)	A	EA	1	1
40	19	PAOZZ	PAOZZ	PAOZZ	PAOZZ	5305-00-211-9344	SCREW, ASSEMBLED P-15121-17 (45722) 69-662-17 (30554)	AC	EA	4	6
40	20	PAOZZ	PAOZZ	PAOZZ	PAOZZ	5935-01-012-3080	CONNECTOR, RECEPTACLE WC596/12-4 (81348)	A	EA	1	1
40	21	PAOZZ	PAOZZ	PAOZZ	PAOZZ	5925-00-686-3298	BREAKER, CIRCUIT MIL-C-5809 (81349) MS25244-7 1/2 (96906)	AC	EA	1	1
40	22	PAOZZ	PAOZZ	PAOZZ	PAOZZ	59254-0-94-4324	BOOT.DUST M5423-14407 (81349) MIL-B-5423-14(81349)	AB	EA	1	2
40	23	PAOZZ	PAOZZ	PAOZZ	PAOZZ	5925-00-966-5836	BREAKER, CIRCUIT 1320SES838-1 (97403) 72-169-2MG6 (74193)	A	EA	1	1
40	24	PAOZZ	PAOZZ	PAOZZ	PAOZZ	5305-00-776-9564	SCREW, ASSEMBLED 42817S8 (23040) ANSI B18.13 (80204)	ABC	EA	1	1
40	25	PAOZZ	PAOZZ	PAOZZ	PAOZZ	5355-00-899-9014	KNOB MS9152S-2K4B (96906)	AC	EA	4	4
40	26	PAOZZ	PAOZZ	PAOZZ	PAOZZ	5930-01-055-9251	SWITCH, ROTARY 72-5011 (30554) 308-431 (44940)	AC	EA	1	1

(1) ILLUSTRATION (a) FIG NO.	(1) ILLUSTRATION (b)	(2) SMR CODE a ARMY	(2) SMR CODE b AIR FORCE	(2) SMR CODE d NAVY	(2) SMR CODE e USMC	(3) NATIONAL STOCK NUMBER	(4) DESCRIPTION REF NUMBER MFR CODE	(4) USABLE ON CODE	(5) U/M	(6) QTY INC IN	(7) USMC QTY PER EQUIP
							GROUP 08 GENERATORCONTROLS AND INSTRUMENTS				
40	27	PAOZZ	PAOZZ	PAOZZ	PAOZZ	5930-00-358-5509	SWITCH, ROTARY MS25002-1 (96906)	A	EA	1	2
40	28	PAOZZ	PAOZZ	PAOZZ	PAOZZ	5905-00-643-5129	RESISTOR, VARIABLE M22-03-00191FD (81349)	ABC	EA	1	1
40	29	PAOZZ	PAOZZ	PAOZZ	PAOZZ	5930-00-259-4646	SWITCH, ROTARY MS25002-2 (96906) 13211E4818 (97403)	AB	EA	1	2
40	30	PAOZZ	PAOZZ	PAOZZ	PAOZZ	5920-00-892-9311	FUSESOLDER, EXTRACTOR FHN26G1 (81349)	A	EA	3	3
40	31	PAOZZ	PAOZZ	PAOZZ	PAOZZ	5920-01-113-2900	FUSE, CARTRIDGE FM09A250V12A (81349)	A	EA	3	3

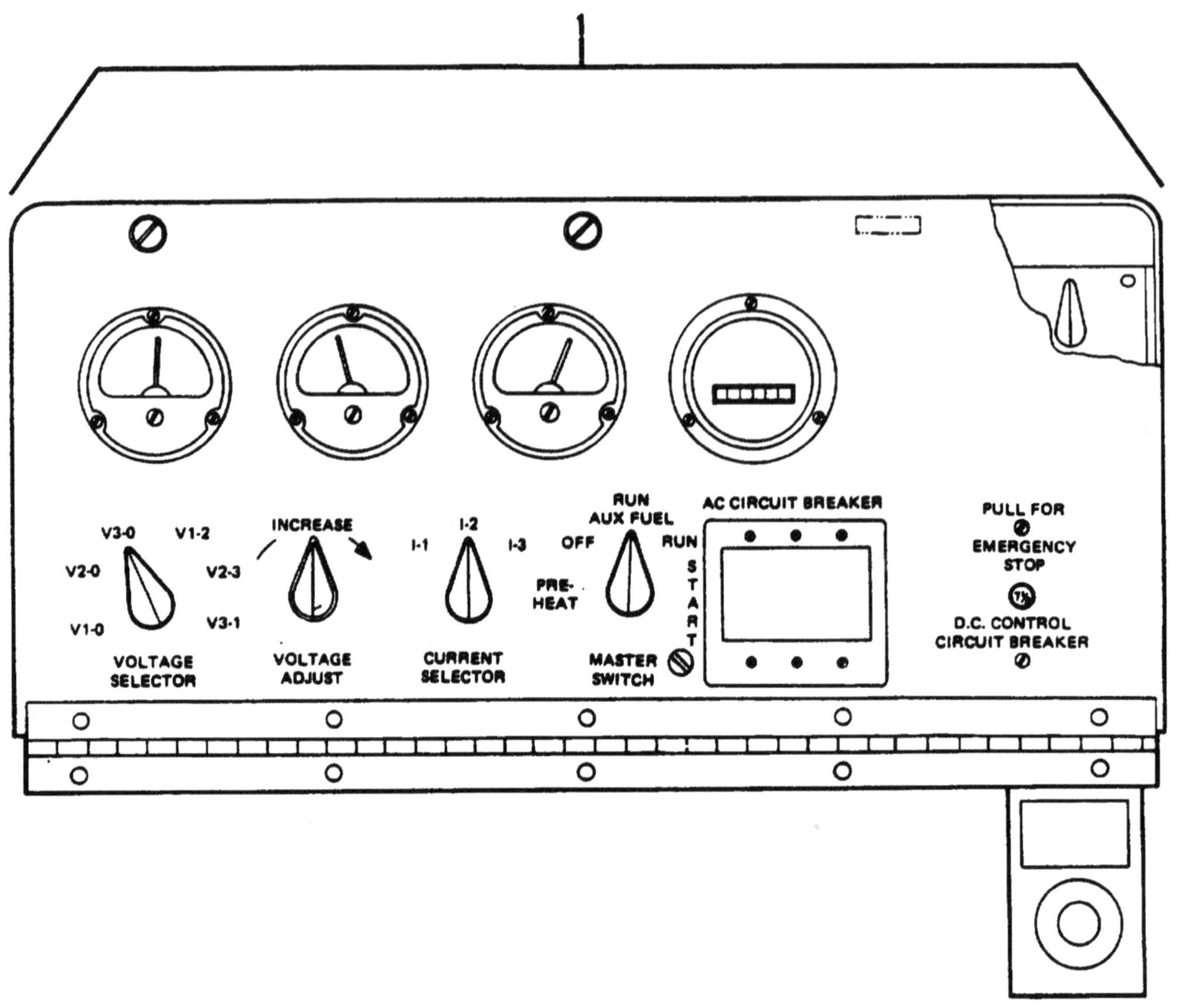

Figure 41. Generator Control 400 Hz, Group 08 (Generator Controls and Instruments)

(1) ILLUSTRA- TION FIG NO	(2) ITEM NO	(3) SMR CODE ARMY	AIR FORCE	NAVY	USMC	QTY USMC	(4) NATIONAL STOCK NUMBER	(5) (6) (7) DESCRIPTIION REF NUMBER	MFR CODE	USABLE ON CODE	U/M	INC IN UNITS	QTY PER EQUIP
								GROUP 08 GENERATOR CONTROLS AND INSTRUMENTS					
41	1	XBOOO	XBOOO	XBOOO	XBOOO			CONTROL,GEN 400 HZ		B	EA	1	1
								300-2963-02	(44940)				
								84-13089-02	(30554)				
								84-13089	(30554)				

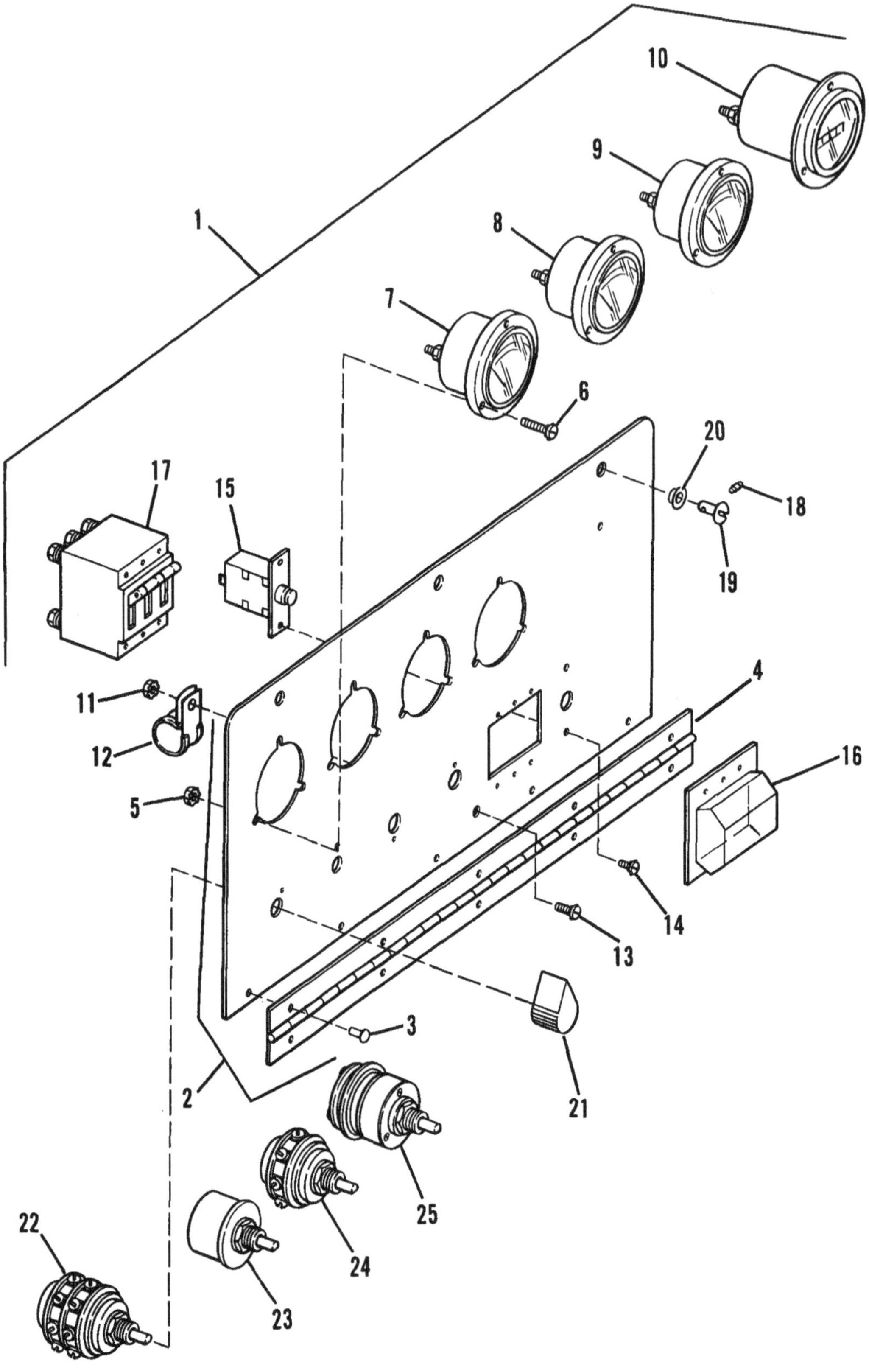

1 2 3 4 5 6 7 8 9 10 11 12 13 14 15 16 17 18 19 20 21 22 23 24 25

Figure 42. Meter Panel 400 Hz, Group 08 (Generator Controls and instruments)

(1) ILLUSTRATION (a) FIG NO.	(b)	(2) SMR CODE a ARMY	b AIR FORCE	d NAVY	e USMC	(3) NATIONAL STOCK NUMBER	(4) DESCRIPTION REF NUMBER MFR CODE	USABLE ON CODE	(5) U/M	(6) QTY INC IN	(7) USMC QTY PER EQUIP
							GROUP 08 GENERATOR CONTROLS AND INSTRUMENTS				
42	1		AOOOO				PANEL, METER 400H 30149325 (44940) 4-13117 (30554)	B	EA	1	1
42	2		XBOZZ				DOOR, METER 319-0188 (44940) 84-13288 (30554)	B	EA	1	1
42	3		PAOZZ			5320-00-395-6523	RIVET M24243/1-D403 (81349)	AB	EA	5	5
42	4		PAOZZ			5340-01-053-7130	HINGE, BUTT MS35823-6C (96906)	B	EA	1	1
42	5		PAOZZ			5310-00-836-3520	NUT, PLAIN 501-040800-00 (78189) 69-561-1 (30554)	BC	EA	12	12
42	6		PAOZZ			5305-00-224-1092	SCREW, ASSEMBLED P15121-5 (45722) 69-662-5 (30554)	BC	EA	12	12
42	7		PAOZZ			6625-00-465-5301	VOLTMETER 13211E6905 (97403)	AB	EA	1	1
42	8		PAOZZ			6625-00-065-8554	METER, FREQUENCY 13211E6992-2 (97403)	B	EA	1	1
42	9		PAOZZ			6625-00-065-5258	METER, ARBITRARY S-10445 (55026) 13211E6919 (97403)	AB	EA	1	1
42	10		PAOZZ			6645-00-089-8842	METER TIME TOTALIZI 15001 04400) M3971/1-5 (81349)	ABC	EA	1	1
42	11		PAOZZ			5310-00-696-5173	NUT, PLAIN 501-250800-00 (78189) 69-561-5 (30554)	ABC	EA	1	1
42	12		PAOZZ			5340-00-702-2848	CLAMP, LOOP MS21333-128 (96906)	ABC	EA	1	1
42	13		PAOZZ			5305-00-776-9564	SCREW, ASSEMBLED 42817S8 (23040) ANSI B18.13 (80204)	AC	EA	1	1
42	14		PAOZZ			5305-00-211-9344	SCREW, ASSEMBLED P15121-17 (45722) 69-662-17 (30554)	BC	EA	2	2

(1) ILLUSTRATION (a) FIG NO.	(b)	(2) SMR CODE a ARMY	b AIR FORCE	d NAVY	e USMC	(3) NATIONAL STOCK NUMBER	(4) DESCRIPTION REF NUMBER MFR CODE	USABLE ON CODE	(5) U/M	(6) QTY INC IN	(7) USMC QTY PER EQUIP
							GROUP 08 GENERATOR CONTROLS AND INSTRUMENTS				
42	15	PAOZZ	PAOZZ	PAOZZ	PAOZZ	5925-00-686-3298	BREAKER, CIRCUIT MS25244-7 1/2 (96906)	BC	EA	1	1
42	16	PAOZZ	PAOZZ	PAOZZ	PAOZZ	5925-00-984-4324	BOOT, DUST M5423-14-07 (81349)	AB	EA	1	1
42	17	PAOZZ	PAOZZ	PAOZZ	PAOZZ	5925-00-682-0742	BREAKER, CIRCUIT 13208E5838-2 (97403) 72-170-1MGL (74193)	B	EA	1	1
42	18	PAOZZ	PAOZZ	PAOZZ	PAOZZ	5315-00-449-2945	PIN, GROOVED 99836 (60119) 69-695 (30554) 1005169 (18876)	ABC	EA	3	3
42	19	PAOZZ	PAOZZ	PAOZZ	PAOZZ		STUD 98292-2-220 (60119) MIL-F-5591 (81349)	ABC	EA	3	3
42	20	PAOZZ	PAOZZ	PAOZZ	PAOZZ	532-00-432-9899	EYELET, METALLIC 406-0371 (44940) 69-766 (30554)' 295901-2 (60119)	BC	EA	3	3
42	21	PAOZZ	PAOZZ	PAOZZ	PAOZZ	5325-00-099-8827	KNOB MS91528-2K4B (96906)	BC	EA	4	4
42	22	PAOZZ	PAOZZ	PAOZZ	PAOZZ	5355-00-899-9014	SWITCH, ROTARY MS25002-2 (96906) 13211E4818 (97403)	AB	EA	1	1
42	23	PAOZZ	PAFZZ	PAFZZ	PAFZZ		RISTOR.VARABLE M22-03-00191FD (81349)	ABC	EA	1	1
42	24	PAOZZ	PAOZZ	PAOZZ	PAOZZ	5930-00-259-4646	SWITCH, ROTARY MS25002-1 (96906)	AB	EA	1	1
42	25	PAOZZ	PAOZZ	PAOZZ	PAOZZ	5905-00-643-5129	KNOB MS91528-2K4B (96906)	BC	EA	1	1
						5930-00-538-5508					
						5355-00-899-9014					

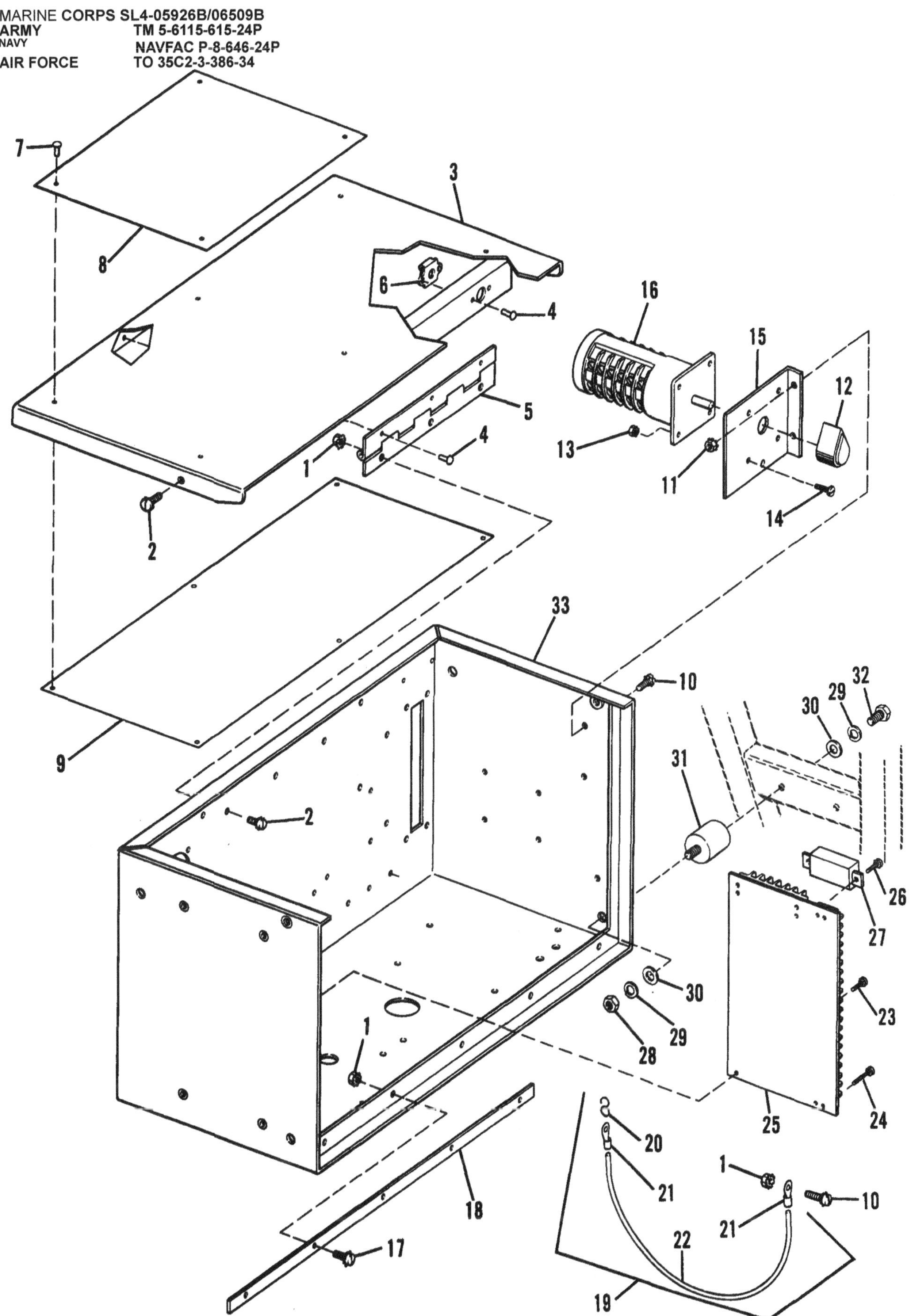

Figure 43. Generator Control Assembly 60,400 Hz, Group 08 (Generator Controls and Instruments) (Sheet 1 of 3)

(1) ILLUSTRATION (a) FIG NO.	(b)	(2) SMR CODE a ARMY	b AIR FORCE	d NAVY	e USMC	(3) NATIONAL STOCK NUMBER	(4) DESCRIPTION REF NUMBER MFR CODE	USABLE ON CODE	(5) U/M	(6) QTY INC IN	(7) USMC QTY PER EQUIP
							GROUP 08 GENERATORCONTROLS AND INSTRUMENTS				
43	1	PAOZZ	PAOZZ	PAOZZ	PAOZZ	5310-00-094-3421	NUT, PLAIN, ASSEMBLED 511-10800-00 (78189) 69-561-4 (30554)	ABC	EA	27	27
43	2	PAOZZ	PAOZZ	PAOZZ	PAOZZ	5305-00-036-6978	SCREW, ASSEMBLED P15121-48 (45722) 69-662-48 (30554)	ABC	EA	7	7
43	3	XBOZZ	XBOZZ	XBOZZ	XBOZZ		COVER, CONTROL BOX 319-0187 (44940) 84-13285 (30554)	AB	EA	1	1
43	4	PAFZZ	PAOZZ	PAOZZ	PAOZZ		RIVET, SOLID MS20426B-6 (96906)	AB	EA	15	15
43	5	PAFZZ	PAOZZ	PAOZZ	PAOZZ	5320-00-165-8772	HINGE, BUTT MS35823-6C (96906)	BA	EA	1	1
43	6	PAOZZ	PAOZZ	PAOZZ	PAOZZ	5340-01-053-7130	RECEPTACLE, TURNLOCK 99947P130 (61864)	AB	EA	3	3
43	7	PAFZZ	PAOZZ	PAOZZ	PAOZZ	5320-00-449-2967	RIVET M24243/1-D403 (81349)	AB	EA	6	6
43	8	MDFZZ	MDOZZ	MDOZZ	MDOZZ	5320-00-395-6523	PLATE, INSTRUCTION 099-2385 (44940) 84-13294 (3054)	AB	EA	1	1
43	9	MDOZZ	MDOZZ	MDOZZ	MDOZZ		PLATE, SCHEM-WIRING O99-2334 (44940) 84-13147 (30554)	AB	EA	1	1
43	10	PAOZZ	PAOZZ	PAOZZ	PAOZZ		SCREW, ASSEMBLED P15121-33 (45722) 69-662-33 (30554)	ABC	EA	11	11
43	11	PAOZZ	PAOZZ	PAOZZ	PAOZZ	5305-00-036-6977	NUT, PLAIN, ASSEMBLED 511-081800-00 (78189) 870-1221 (44940) 69-561-3 (30554)	ABC	EA	20	20
43	12	XAOZZ	XAOZZ	XAOZZ	XAOZZ	5310-00-052-3632	KNOB, SWITCH B18C00590E (10983)	AB	EA	1	1
43	13	PAOZZ	PAOZZ	PAOZZ	PAOZZ		NUT, PLAIN, ASSEMBLED 511-061800-00 (78189) 69-561-2 (30554)	AB	EA	10	10
						5310-00-063-7360					

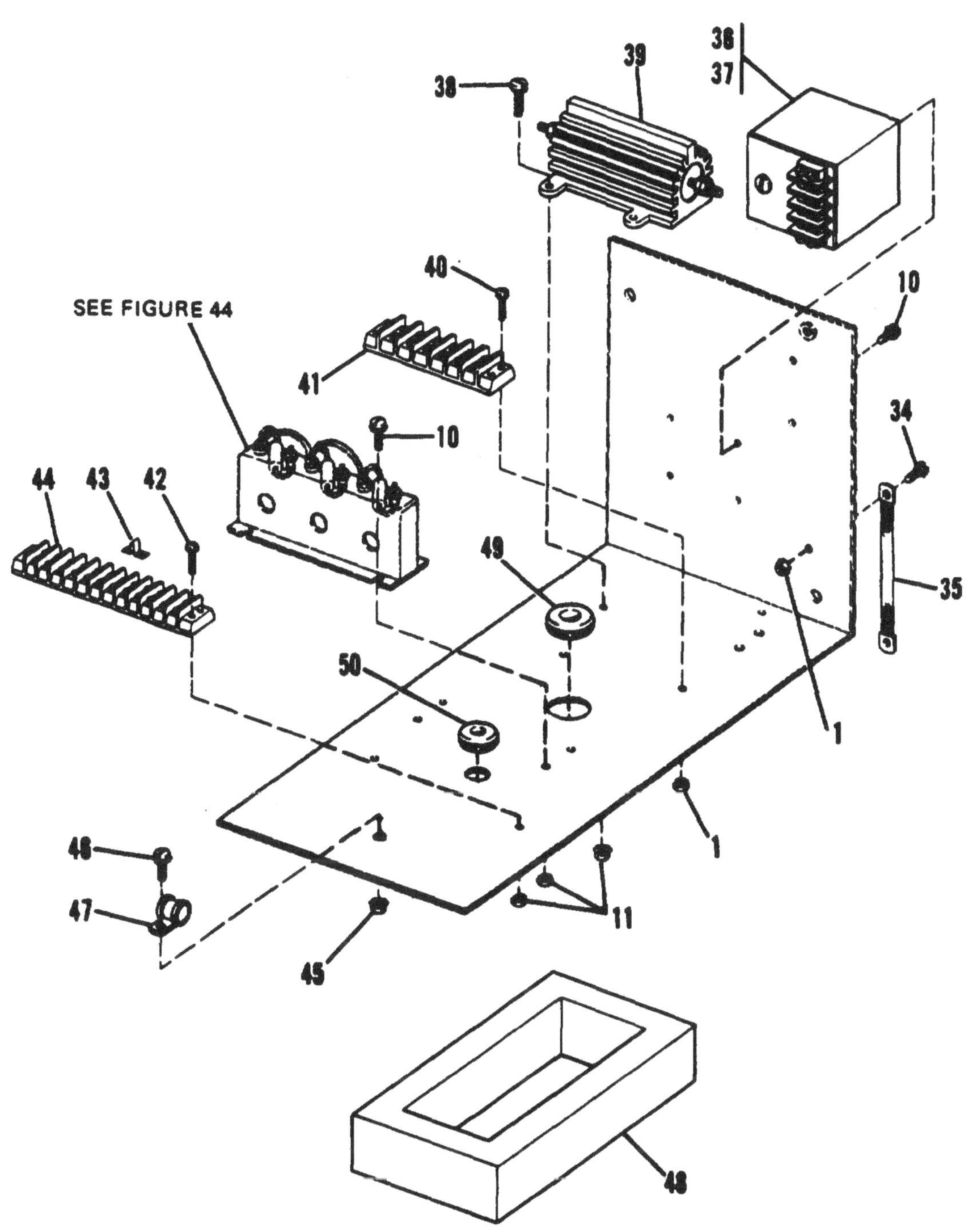

Figure 43. Generator Control Assembly 60, 400 Hz, Group 08 (Generator Controls and Instruments) (Sheet 2 of 3)

(1) ILLUSTRATION (a) FIG NO.	(b)	(2) SMR CODE a ARMY	b AIR FORCE	d NAVY	e USMC	(3) NATIONAL STOCK NUMBER	(4) DESCRIPTION REF NUMBER MFR CODE / USABLE ON CODE	(5) U/M	(6) QTY INC IN	(7) USMC QTY PER EQUIP
							GROUP 08 GENERATOR CONTROLS AND INSTRUMENTS			
43	14	PAOZZ	PAOZZ	PAOZZ	PAOZZ	5305-00-036-6972	SCREW, ASSEMBLED .... ABC P15121-20 (45722) 69-662-20 (30554)	EA 4	4	
43	15	XBOZZ	XBOZZ	XBOZZ	XBOZZ		BRACKET....... AB 319-0193 (44940) 84-13284 (30554)	EA 1	1	
43	16	PAOZZ	PAOZZ	PAOZZ	PAOZZ	5930-00-659-2672	SWITCH.ROTARY....... C 13213E4091 (97403) PR20-0112A4-1 (82121)	EA 1	1	
43	17	PAOZZ	PAOZZ	PAOZZ	PAOZZ		SCREW, ASSEMBLED WASHER .... ABC 303737 (32195)			
43	18	XBOZZ	XBOZZ	XBOZZ	XBOZZ	5305-00-177-5778	PLATE., SPACER....... C 301-9483 (44940) 84-13148 (30554)	EA	5	5
43	19	AFOZZ	AFOZZ	AFOZZ	AFOZZ		HOLDER, METER PANEL....... ABC 406-0853 (44940) 84-13069 (30554)	EA 1	1	
43	20	PAOZZ	PAOZZ	PAOZZ	PAOZZ		HOOK, CHAINS....... ABC MS87006-13 (96906)	EA 1	1	
43	21	PAOZZ	PAOZU	PAOZZ	PAOZZ	4030-00-780-9350	TERMINAL, LUG .... ABC MS20659-144 (96906)	EA 1	1	
43	22	PAOZZ	PAOZZ	PAOZZ	PAOZZ	5940-00-115-2677	ROPE, NYLON .... ABC 8198810 (19200)			
43	23	PAOZZ	PAOZZ	PAOZZ	PAOZZ	4020-00-523-9641	SCREW, ASSEMBLED .... ABC 69-662-18 (30554) P15121-18 (45722)	EA 2	2	
43	24	PAOZZ	PAOZZ	PAOZZ	PAOZZ	5305-00-036-6968	SCREW, ASSEMBLED .... ABC P15121-21 (45722) 69-662-21 (30554)	IN 9	9	
43	25	PAOFF	PAOFF	PAOFF	PAOFF	5305-00-036-6976	CIRCUIT CARD ASSY....... AB 300-2953 (44940) 84-13178 (30554)	EA 5	5	
43	26	PAFZZ	PAFZZ	PAFZZ	PAFZZ		SCREW, ASSEMBLED .... AB P15121-2 (45722) 69-662-2 (30554)	EA 5	5	
43	27	PAFZZ	PAFZZ	PAFZZ	PAFZZ	5998-01-281-0071	RELAY, ELEC .... AB M5757/23-003 (81349)	EA	1	1
43	28	PAOZZ	PAOZZ	PAOZZ	PAOZZ	5305-00-36-6970	NUT, PLAIN, HEXAGON .... ABC MS51967-6 (96906)	EA 8	8	
						5945-00-435-1833		EA 4	4	
								EA	4	

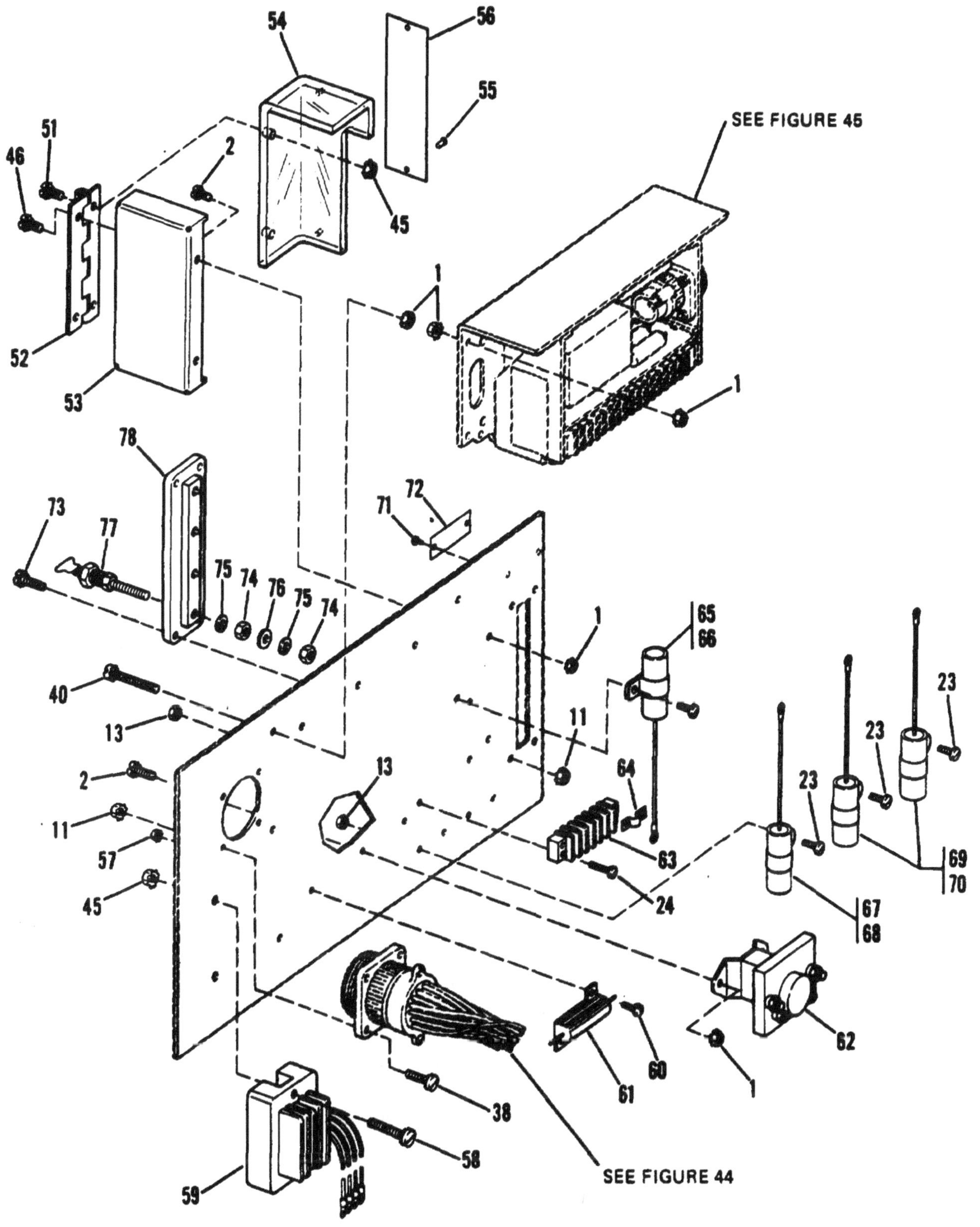

Figure 43. Generator Control Assembly 60, 400 Hz, Group 08 (Generator Controls and Instruments) (Sheet 3 of 3)

(1) ILLUS-TRA-TION		(2) SMR CODE				(3)	(4)	(5)	(6)	(7)
(a) FIG NO.	(b)	a ARMY	b AIR FORCE	d NAVY	e USMC	NATIONAL STOCK NUMBER	DESCRIPTION USABLE ON CODE REF NUMBER MFR CODE	U/M	QTY INC IN	USMC QTY PER EQUIP
							GROUP 08 GENERATOR CONTROLS AND INSTRUMENTS			
43	29	PAOZZ	PAOZZ	PAOZZ	PAOZZ	5310-00-407-9566	WASHER, LOCK .... ABC MS35338-45 (9906) 850-1045 (44191)	EA	8	8
43	30	PAOZZ	PAOZZ	PAOZZ	PAOZZ	5310-00-081-4219	WASHER, FLAT .... ABC MS27183-12 (96906)	EA	8	8
43	31	PAOZZ	PAOZZ	PAOZZ	PAOZZ	5340-01-275-3534	MOUNT, RESILIENT .... ABC 29031 (81860) 84-13015 (30554)	EA	4	4
43	32	PAOZZ	PAOZZ	PAOZZ	PAOZZ		BOLT, MACHINE.......... ABC MS90725-31 (96906)	EA	4	4
43	33	XBOZZ	XBOZZ	XBOZZ	XBOZZ	5306-00-225-8496	BOX, CONTROL..........AB 301-9285 (44940) 84-13162 (30554)	EA	1	1
43	34	PAOZZ	PAOZZ	PAOZZ	PAOZZ		SCREW, ASSEMBLED .... ABC P-15121-50 (45722) 69-662-50 (30554)	EA	1	1
43	35	PAOZZ	PAOZZ	PAOZZ	PAOZZ	5305-00-036-6906	STRAP, GROUND..........AB 72-5006-1 (30554)	EA	1	1
43	36	PAOZZ	PAOZZ	PAOZZ	PAOZZ	5340-01-233-8305	CONVERTER, CONTROL.......... A 13211E6901 (97403) 4431 (13483)	EA	1	1
43	37	PAOZZ	PAOZZ	PAOZZ	PAOZZ	6115-00-940-0175	CONVERTER, FREQUENCY.......... B 13212E8933 (97403) 11206 (12670) 20-1085-2 (03776)	EA	1	1
43	38	PAOZZ	PAOZZ	PAOZZ	PAOZZ	6115-659-2786	SCREW, ASSEMBLED .... ABC P15121-35 (45722) 69-662-35 (30554)	EA	7	7
43	39	PAOZZ	PAOZZ	PAOZZ	PAOZZ		RESISTOR, FIXED, WIRE.......... ABC RE80G1R00 (81349)	EA	1	1
43	40	PAOZZ	PAOZZ	PAOZZ	PAOZZ	5305-00-038-3103	SCREW, ASSEMBLED ....AB P15121-52 (45722) 69-662-52 (30554)	EA	6	6
43	41	PAOZZ	PAOZZ	PAOZZ	PAOZZ	5905-00-883-8431	TERMINAL BOARD..........AB 39TB6F (81349) 22006 (26405)	EA	1	1
43	42	PAOZZ	PAOZZ	PAOZZ	PAOZZ	5305-00-224-1093	SCREW, ASSEMBLED .... ABC P15121-37 (45722) 69-662-37 (30554)	EA	2	2
						5940-00-983-6101				

(1) ILLUSTRATION (a) FIG NO.	(1) (b)	(2) SMR CODE a ARMY	b AIR FORCE	d NAVY	e USMC	(3) NATIONAL STOCK NUMBER	(4) DESCRIPTION / REF NUMBER MFR CODE / USABLE ON CODE	(5) U/M	(6) QTY INC IN	(7) USMC QTY PER EQUIP
							GROUP 08 GENERATOR CONTROLS AND INSTRUMENTS			
43	43	PAOZZ	PAOZZ	PAOZZ	PAOZZ	6150-00-519-2714	BUS, CONDUCTOR ...... ABC M55164/28A-TBJB (81349)	EA	3	3
43	44	PAOZZ	PAOZZ	PAOZZ	PAOZZ	5940-00-983-6089	TERMINAL BOARD...... ABC 38TB12 (81349)	EA	1	1
43	45	PAOZZ	PAOZZ	PAOZZ	PAOZZ	5310-00-096-5173	NUT, PLAIN, ASSEMBLED ...... ABC 501-250800-00 (78189) 69-561-5 (30554)	EA	7	7
43	46	PAOZZ	PAOZZ	PAOZZ	PAOZZ		SCREW, ASSEMBLED ...... ABC 42817S8 (23040) ANSI B18.13 (80204)	EA	5	5
43	47	PAOZZ	PAOZZ	PAOZZ	PAOZZ	5305-00-776-9564	CLAP, LOOP ...... ABC MS21333-128 (96906)	EA	1	1
43	48	XBOZZ	XBOZZ	XBOZZ	XBOZZ	5340-00-702-2848	DUCT AIR SPONGE...... ABC 134-4543 (44940) 84-13181 (30554)	EA	1	1
43	49	PAOZZ	PAOZZ	PAOZZ	PAOZZ		GROMMET...... ABC MS35489-123 (96906)	EA	1	1
43	50	PAOZZ	PAOZZ	PAOZZ	PAOZZ	5325-00-754-2187	GROMMET...... ABC 2860 (70485) 205K1114 (94990)	EA	1	1
43	51	PAOZZ	PAOZZ	PAOZZ	PAOZZ	5325-00-185-0003	SCREW, ASSEMBLED ...... ABC P-15121-78 (45722) 69-662-78 (30554)	EA	2	2
43	52	PAOZZ	PAOZZ	PAOZZ	PAOZZ	5305-01-078-5064	HINGE, BUTT...... ABC 406-0581 (44940) 84-13190 (30554)	EA	1	1
43	53	XBOZZ	XBOZZ	XBOZZ	XBOZZ	5340-01-281-5270	BRACKET, MOUNTING ...... ABC 319-0194 (44940) 84-13293 (30554)	EA	1	1
43	54	PAOZZ	PAOZZ	PAOZZ	PAOZZ		COVER, ELECTRICAL......AC 332-2767 (44940) 84-13171 (30554)	EA	1	1
43	55	PAFZZ	PAO7Z	PAOZZ	PAOZZ		RIVET...... AB EA M24243/1-D405 (81349)	2	2	
43	56	MDFZZ	MDOZZ	MDOZZ	MDOZZ	5935-01-280-1177	PLATE, TERMINAL ID......AB 099-2381 (44940) 84-13295 (30554)	EA	1	1
						5320-00-971-				

(1) ILLUSTRATION (a) FIG NO.	(b)	(2) SMR CODE a ARMY	b AIR FORCE	d NAVY	e USMC	(3) NATIONAL STOCK NUMBER	(4) DESCRIPTION REF NUMBER MFR CODE	USABLE ON CODE	(5) U/M	(6) QTY INC IN	(7) USMC QTY PER EQUIP
							GROUP 08 GENERATOR CONTROLS AND INSTRUMENTS				
43	57	PAOZZ	PAOZZ	PAOZZ	PAOZZ	5310-00-836-3520	NUT, PLAIN, ASSEMBLED 501-040800-00 69-561-1 (30554)	ABC	EA 2	2	
43	58	PAOZZ	PAOZZ	PAOZZ	PAOZZ	5305-01-114-5801	SCREW, ASSEMBLED P-15121-82 (45722) 69-662-82 (30554)	ABC	EA 2	2	
43	59	PAOZZ	PAOZZ	PAOZZ	PAOZZ	2920-01-282-8522	REGULATOR, ENGINE 12720 (2N1I4) 84-13183 (30554)	ABC	EA 1	1	
43	60	PAOZZ	PAOZZ	PAOZZ	PAOZZ	5305-00-038-3089	SCREW, ASSEMBLED P15121-3 (45722) 69-662-3 (30554)	ABC	EA 2	2	
43	61	PAOZZ	PAOZZ	PAOZZ	PAOZZ	5905-00-139-1989	RESISTOR, FIXED, WIRE RER75F4R02R (1349)	AB	EA 1	1	
43	62	PAOZZ	PAOZZ	PAOZZ	PAOZZ	5945-00-686-6877	RELAY MS2416-D1 (96906)	ABC	EA 1	1	
43	63	PAOZZ	PAOZZ	PAOZZ	PAOZZ	5940-00-983-6048	BOARD, TERMNAL 37TB7 (81349)	AB	EA 1	1	
43	64	PAOZZ	PAOZZ	PAOZZ	PAOZZ	615 0632-7234	BUS, CONDUCTOR 600J (83330) 12718025-2 (19200)	AB	EA 1	1	
43	65	PAOZZ	PAOZZ	PAOZZ	PAOZZ	5910-01-279-0003	CAPACITOR, FIXED 84-13310-03 (30554)	A	EA 1	1	
43	66	PAOZZ	PAOZZ	PAOZZ	PAOZZ	5910-01-280-0754	CAPACITOR, FIXED 312-0260-02 (44940) 84-13311-02 (30554)	ABC	EA 1	1	
43	67	PAOZZ	PAOZZ	PAOZZ	PAOZZ	5910-01-283-9069	CAPACITOR, FIXED 84-13310-02 (30554)	A	EA 1	1	
43	68	PAOZZ	PAOZZ	PAOZZ	PAOZZ	5910-01-280-0754	CAPACITOR, FIXED 312-0260-02 (44940) 84-13311-02 (30554)	B	EA 1	1	
43	69	PAOZZ	PAOZZ	PAOZZ	PAOZZ	5910-01-283-6881	CAPACITOR, FIXED 84-13310-01 (30554)	A	EA 2	2	
43	70	PAOZZ	PAOZZ	PAOZZ	PAOZZ	5910-01-280-	CAPACITOR, FIXED 312-0260-02 (44940) 84-13311-02 (30554)	B	EA 2	2	

(1) ILLUSTRATION (a) FIG NO.	(b)	(2) SMR CODE a ARMY	b AIR FORCE	d NAVY	e USMC	(3) NATIONAL STOCK NUMBER	(4) DESCRIPTION USABLE ON CODE / REF NUMBER MFR CODE	(5) U/M	(6) QTY INC IN	(7) USMC QTY PER EQUIP
							GROUP 08 GENERATOR CONTROLS AND INSTRUMENTS			
43	71	PAFZZ	PAOZZ	PAOZZ	PAOZZ	5320-01-004-023	RIVET, BLIND ....AB M24243/1-D402 (81349)	EA	2	2
43	72	MDFZZ	MDOZZ	MDOZZ	MDOZZ		PLATE, ID.... A 13211E6888-2 (97403)	EA	1	1
43	73	PAOZZ	PAOZZ	PAOZZ	PAOZZ	5305-00-958-5477	SCREW .... ABC MS35190-254 (96906)	EA	4	4
43	74	PAOZZ	PAOZZ	PAOZZ	PAOZZ	5310-01-301-1962	NUT, HEX.... ABC MS35649-2255N (96906)	EA	8	8
43	75	PAOZZ	PAOZZ	PAOZZ	PAOZZ	5310-00-1-8970	WASHER, LOCK ....AB MS35338-101 (96906)	EA	8	8
43	76	PAOZZ	PAOZZ	PAOZZ	PAOZZ	5310-00-950-0440	WASHER, FLAT .... ABC MS15795-409 (96906)	EA	4	4
43	77	PAOZZ	PAOZZ	PAOZZ	PAOZZ	5940-00-958-0349	TERMINAL, STUD....AB 13208E5820-2 (97403) DG3M09F-S1-RPC (82168)	EA	4	4
43	78	PAOZZ	PAOZZ	PAOZZ	PAOZZ	5940-00-477-9967	BOARD.TERMINAL....AB 13215E1942 (97403)	EA	1	1

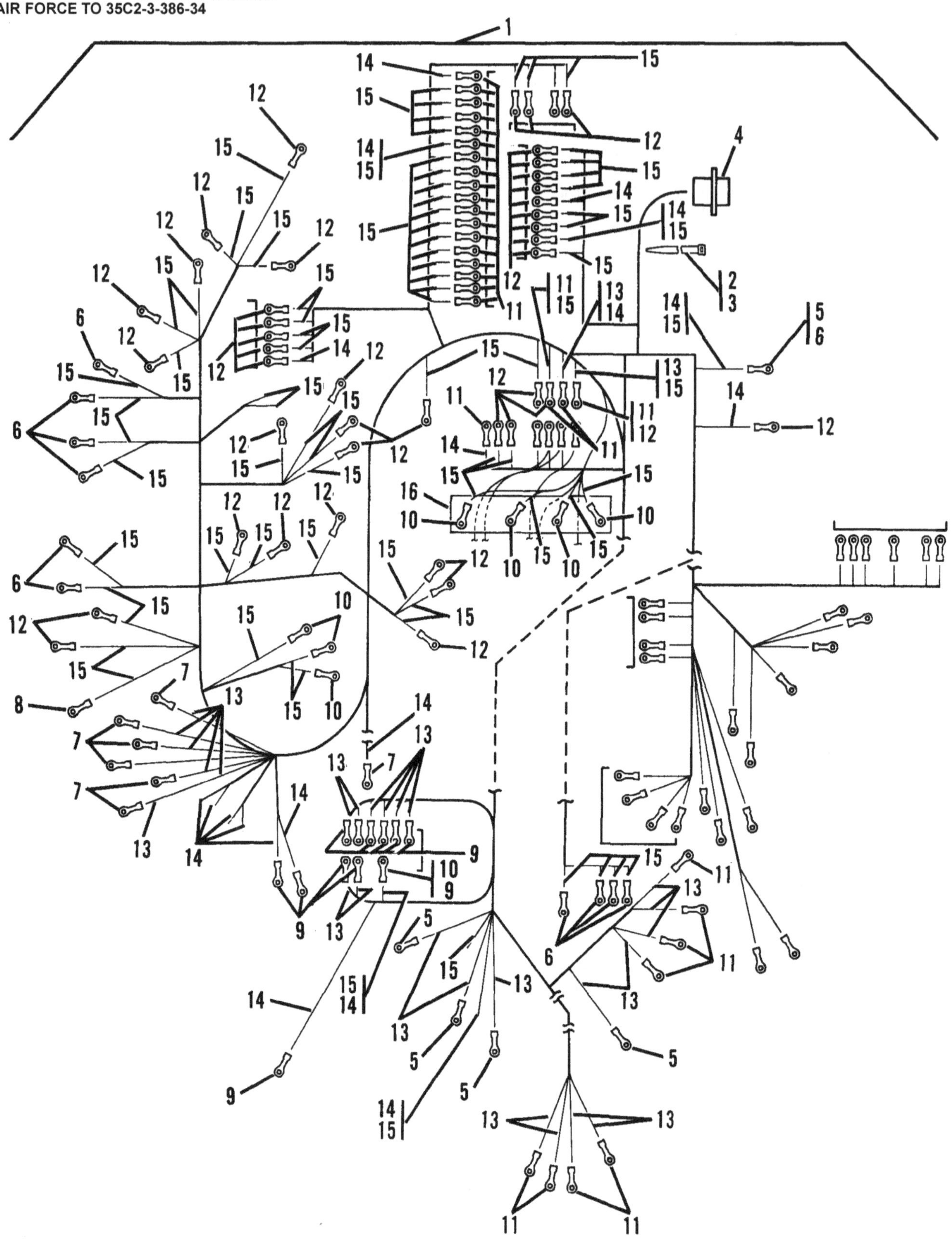

Figure 44. Control Wire Harness 60,400 Hz, Group 08 (Generator Controls and Instruments) (Sheet 1 of 2)

(1) ILLUS-TRA-TION (a) FIG NO.	(b)	(2) SMR CODE a ARMY	b AIR FORCE	d NAVY	e USMC	(3) NATIONAL STOCK NUMBER	(4) DESCRIPTION / USABLE ON CODE / REF NUMBER MFR CODE	(5) U/M	(6) QTY INC IN	(7) USMC QTY PER EQUIP
							GROUP 08 GENERATOR CONTROLS AND INSTRUMENTS			
44	1	AOOOOAOOOO		AOOOO			HARNESS ASSY....AB 338-1897 (44940) 84-13154 (30554)	EA	1	1
44	2	PAOZZ	PAOZZ	PAOZZ		5975-00-074-2072	STRAP, TIE DOWN....AB MS3367-1-9 (96906)	EA	40	40
44	3	PAOZZ	PAOZZ	PAOZZ			STRAP, TIE DOWN....AB MS3367-5-9 (96906)	EA	80	80
						5975-00-111-3208				
44	4	PAOZZ	PAOZZ	PAOZZ			CONNECTOR, RECEPT ....AB MS3450W32-7P (96906)	EA	1	1
						5935-01-219-4205				
44	5	PAOZZ	PAOZZ	PAOZZ			TERMINAL, LUG ....AB MS25036-157 (96906)	EA	19	19
44	6	PAOZZ	PAOZZ	PAOZZ		5940-00-143-4777	TERMINAL, LUG ....AB MS25036-154 (96906)	EA	15	15
44	7	PAOZZ	PAOZZ	PAOZZ		5940-00-230-0515	TERMINAL, LUG ....AB MS25036-112 (96906)	EA	5	5
44	8	PAOZPAOZZ		PAOZZ			TERMINAL, LUG ....AB MS25036-IOS (96906)	EA	1	1
						5940-00-143-4794				
44	9	PAOZZ	PAOZZ	PAOZZ			TERMINAL, LUG ....AB MS2036-156 (96906)	EA	6	6
						5940-00-143-4780				
44	10	PAOZZ	PAOZZ	PAOZZ			TERMINAL, LUG ....AB MS25036-153 (96906)	EA	13	13
						5940-00-143-4775				
44	11	PAOZZ	PAOZZ	PAOZZ			TERMINAL, LUG ....AB MS17143-15 (96906)	EA	26	26
44	12	PAOZZ	PAOZZ	PAOZZ			TERMINAL, LUG ....AB MS25036-106 (96906)	EA 101		101
						5940-00-143-4774				
44	13	PAOZZ	PAOZZ	PAOZZ			WIRE, ELECTRICAL ....AB M50020-10-9 (81349)	FT	16	16
						5940-00-836-0360				
44	14	PAOZZ	PAOZZ	PAOZZ			WIRE, ELECTRICAL ....AB M50OS62-12-9 (81349)	FT	16	16
44	15	PAOZZ	PAOZZ	PAOZZ		5940-00-283-5280	WIRE, ELECTRICAL ....AB M5O6I3-16-9 (81349)	FT	16	16
44	16	PAOZZ	PAOZZ	PAOZZ		6145-00-578-7513	TRANSFORMER, CURRENT....AB 13213E412S (97403)	EA	1	1
						6145-00-578-7514				
						6145-00-655-2562				

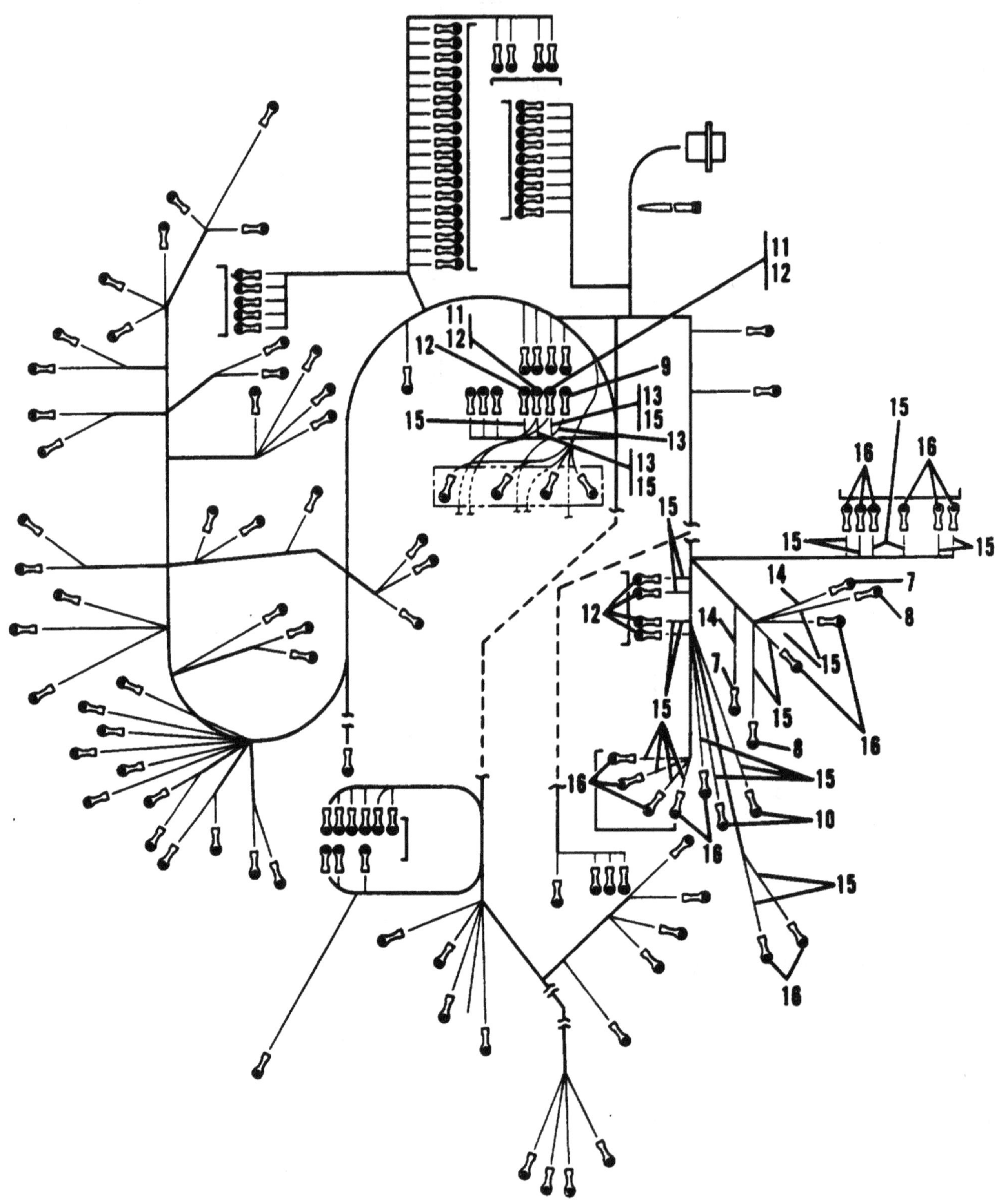

Figure 44. Control Wire Harness 60, 400 Hz, Group 08 (Generator Controls and Instruments) (Sheet 2 of 2)

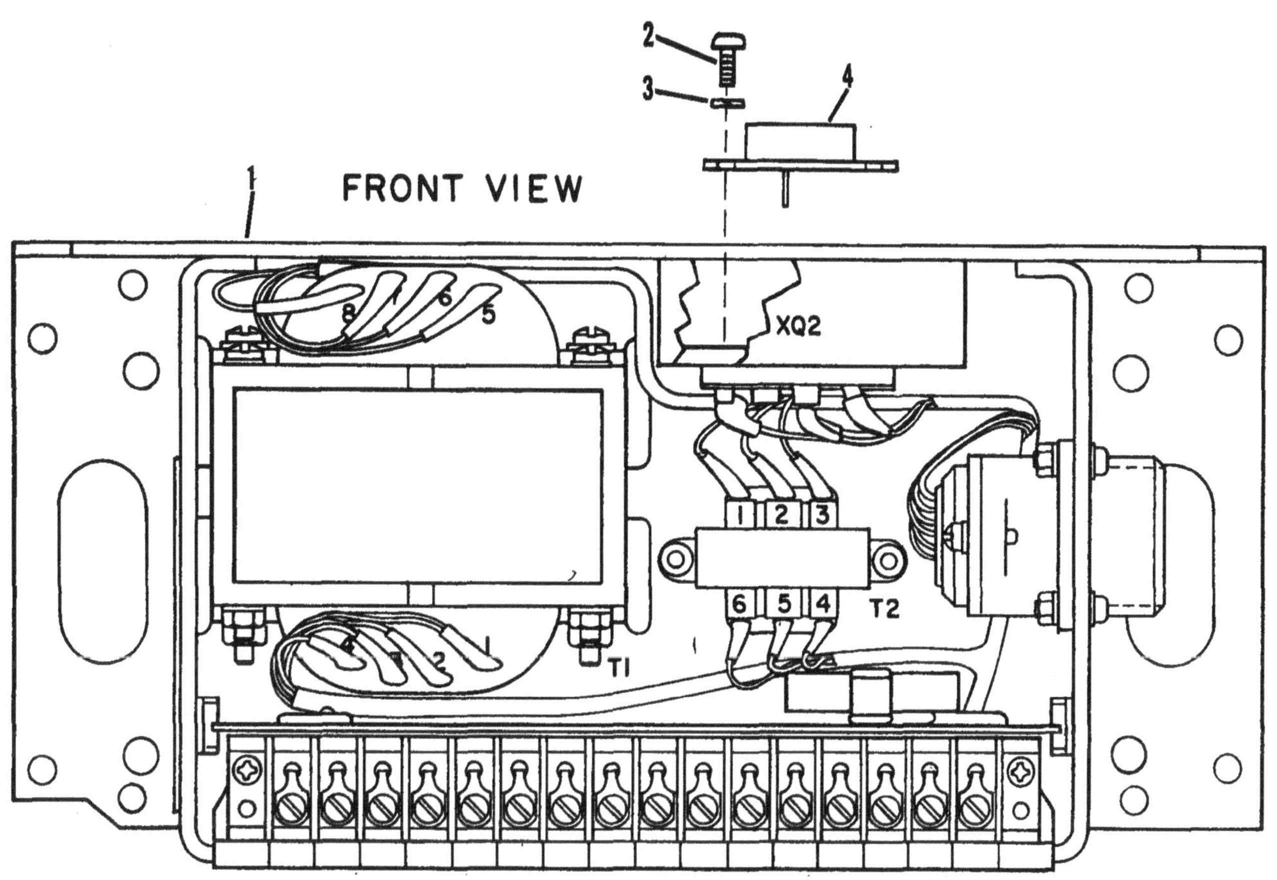

Figure 45. Voltage Regulator, Group 08 (Generator Controls and instruments)

(1) ILLUSTRATION (a) FIG NO.	(1) ILLUSTRATION (b)	(2) SMR CODE a ARMY	(2) SMR CODE b AIR FORCE	(2) SMR CODE d NAVY	(2) SMR CODE e USMC	(3) NATIONAL STOCK NUMBER	(4) DESCRIPTION / REF NUMBER MFR CODE	(4) USABLE ON CODE	(5) U/M	(6) QTY INC IN	(7) USMC QTY PER EQUIP
							GROUP 08 GENERATOR CONTROLS AND INSTRUMENTS				
45	1	PAFFF	PAFFF	PAFFF	PAFFF	6110-00-764-7621	REGULATOR, VOLTAGE 13219E0813 (97403)	ABC	EA	1	1
45	2	PAFZZ	PAFZZ	PAFZZ	PAFZZ	5305-00-889-3001	SCREW, MACHINE MS35206-231 (96906)	ABC	EA	2	2
45	3	PAFZZ	PAFZZ	PAFZZ	PAFZZ	5310-00-0454007	WASHER, LOCK MS3533841 (96906)	ABC	EA	2	2
45	4	PAFZZ	PAFZZ	PAFZZ	PAFZZ	5961-00-712-5578	TRANSISTOR MIL-S-19500-370 (81349) 2N3442 (81349)	ABC	EA	1	1

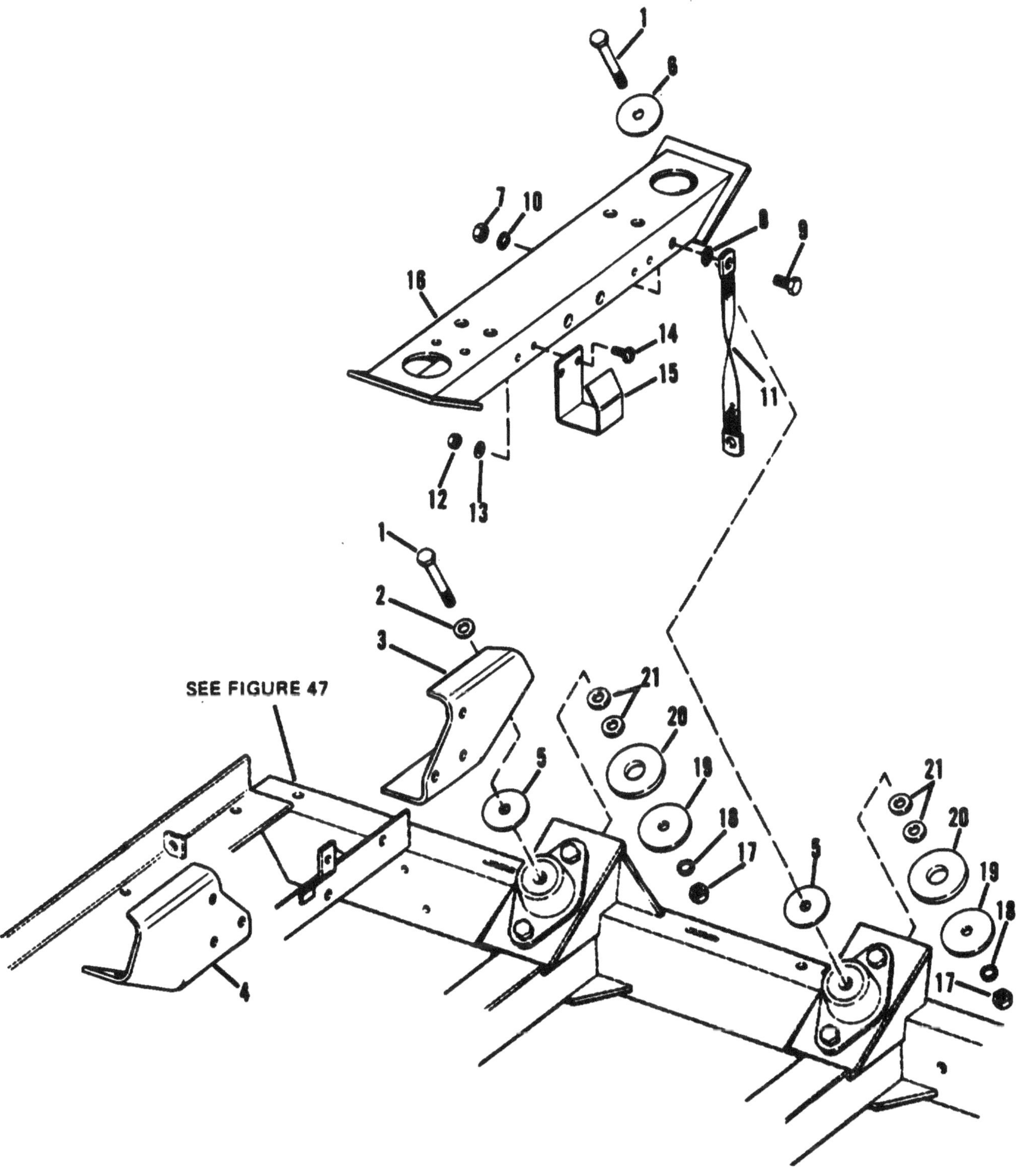

Figure 46. Engine Bracket and Generator Mount, Group 09 (Skid Base)

(1) ILLUSTRATION (a) FIG NO.	(1) ILLUSTRATION (b)	(2) SMR CODE a ARMY	(2) SMR CODE b AIR FORCE	(2) SMR CODE d NAVY	(2) SMR CODE e USMC	(3) NATIONAL STOCK NUMBER	(4) DESCRIPTION / REF NUMBER MFR CODE	(4) USABLE ON CODE	(5) U/M	(6) QTY INC IN	(7) USMC QTY PER EQUIP
							GROUP 09 SKID BASE				
46	1	PAOZZ	PAOZZ	PAOZZ	PAOZZ	5305-00-269-3219	SCREW, CAP, HEXAGON MS90725-69 (96906)	ABC	EA	4	4
46	2	PAOZZ	PAOZZ	PAOZZ	PAOZZ	5310-00-229-4677	WASHER, FLAT MS2440-6 (96906)	ABC	EA	2	3
46	3	PAOZZ	PAOZZ	PAOZZ	PAOZZ	5340-01-305-3414	BRACKET, MOUNTING 403-2358 (4494) 84-13027 C054)	ABC	EA	1	1
46	4	XBOZZ	XBOZZ	XBOZZ	XBOZZ		BRKT ENG, START SIDE 403-2359 (44940) 84-13C28 (30554)	ABC	EA	1	1
46	5	PAOZZ	PAOZZ	PAOZZ	PAOZZ		WASHER, FLAT 5261023 (44940) 84-13043 (30554)	ABC	EA	4	4
46	6	PAOZZ	PAOZZ	PAOZZ	PAOZZ	5310-01-275-3324	WASHER, FLAT 526-0240, (44940) 72-5217-1 (30554)	ABC	EA	2	2
46	7	PAOZZ	PAOZZ	PAOZZ	PAOZZ	5310-01-049-4077	NUT, PLAIN, HEXAGON MS51967-9 (96906)	ABC	EA	2	7
46	8	PAOZZ	PAOZZ	PAOZZ	PAOZZ	5310-00-761-0654	WASHER, LOCK MS35335-35 (96906)	ABC	EA	1	1
46	9	PAOZZ	PAOZZ	PAOZZ	PAOZZ		SCREW, CAP MS18154-60 (96906)	ABC	EA	2	2
46	10	PAOZZ	PAOZZ	PAOZZ	PAOZZ	5310-00-627-6128	WASHER, FLAT MS254406 (96906)	ABC	EA	1	1
46	11	PAOZZ	PAOZZ	PAOZZ	PAOZZ	5305-00-942-2196	LEAD, ELECTRICAL 33740096 (44940) 72-5OO62 (30554)	ABC	EA	1	1
46	12	PAOZZ	PAOZZ	PAOZZ	PAOZZ	5310-00-229-4677	NUT, PLAIN, ASSEMBLED 501-250800-00 (78189) 69-561-5 (30554)	ABC	EA	4	4
46	13	PAOZZ	PAOZZ	PAOZZ	PAOZZ	6150-01-051-0145	WASHER, FLAT MS27183-10 (96906)	ABC	EA	4	4
46	14	PAOZZ	PAOZZ	PAOZZ	PAOZZ	5310-00-696-5173	SCREW, ASSEMBLED P - 15121-79 (45722) 69-662-79 (30554)	ABC	EA	4	4
46	15	XBOZZ	XBOZZ	XBOZZ	XBOZZ		BRACKET.HOSE 403-2363 (44940) 84-13033 (30554)	ABC	EA	2	2
						5310-00-809-4058					

(1) ILLUSTRATION (a) FIG NO.	(1) ILLUSTRATION (b) ITEM NO.	(2) SMR CODE a	(2) SMR CODE b AIR	(2) SMR CODE d	(2) SMR CODE e USMC	(3) NATIONAL STOCK NUMBER	(4) DESCRIPTION REF NUMBER MFR CODE / USABLE ON CODE	(5) U/M	(6) QTY INC IN	(7)
							GROUP 09 SKID BASE			
46	16		XBOZZ		XBOZZ		BRACKET MTG GEN .... ABC 403-2360 (44940) 84-13032 (30554)	EA	1	1
46	17		PAOZZ		PAOZZ	5310-00-761-4654	NUT, PLAIN, HEXAGON .... ABC MS51967-9 (96906)	EA	5	5
46	18		PAOZZ		PAOZZ	5310-00-637-9541	WASHER, LOCK .... ABC MS35338-46 (96906)	EA	4	4
46	19		PAOZZ		PAOZZ	5310-01-275-3325	WASHER, FLAT .... ABC 526-1024 (44940) 84-13044 (30554)	EA	4	4
46	20		PAOZZ		PAOZZ	5310-01-275-3326	WASHER, FLAT .... ABC 402-0573 (44940) 84-13041 (30554)	EA	4	4
46	21		PAOZZ		PAOZZ	5310-01-010-2261	WASHER, FLAT .... ABC 526-0174 (44940)	EA	8	8

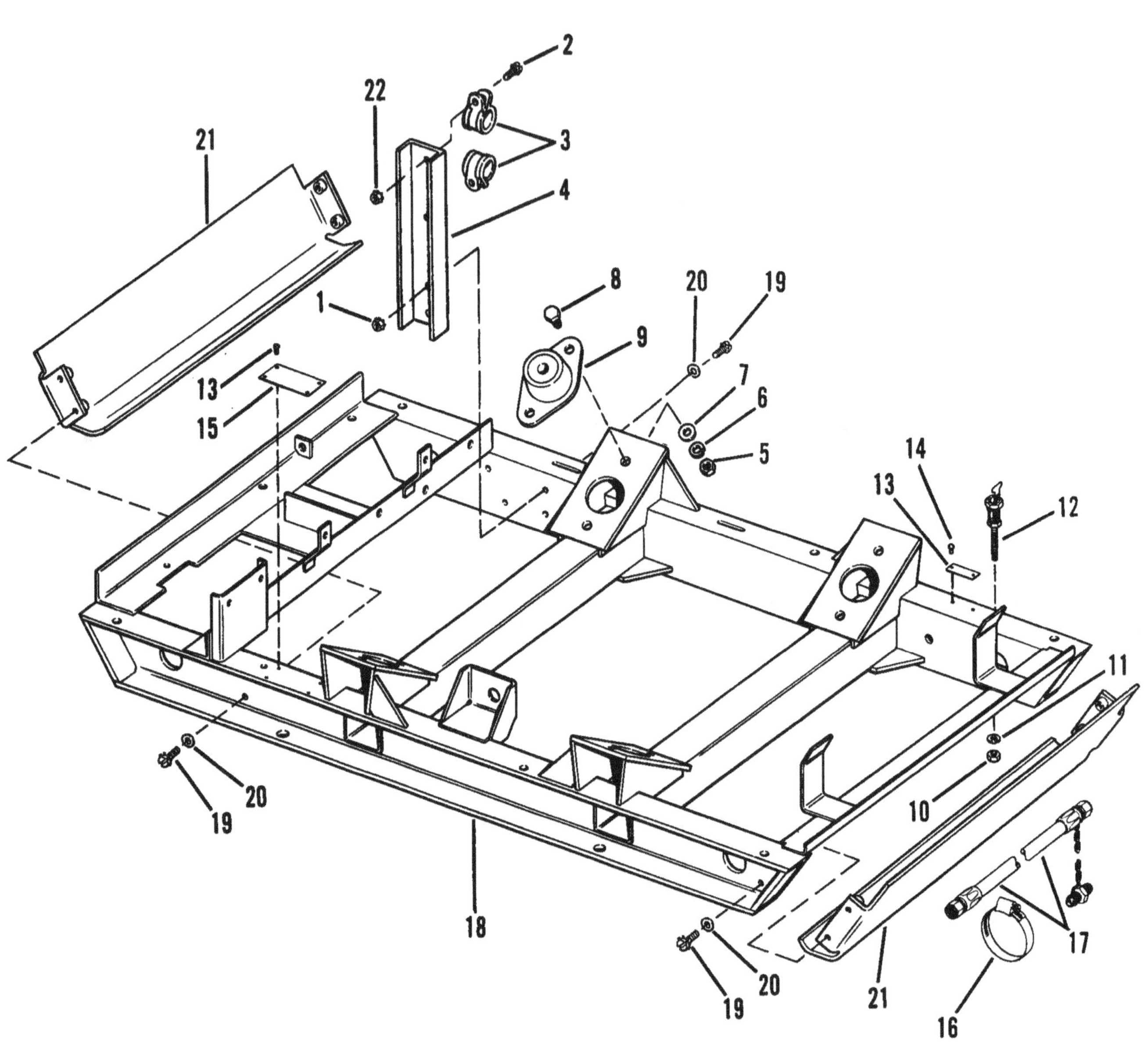

Figure 47. Skid Base and Engine Mount, Group 09 (Skid Base)

(1) ILLUSTRATION (a) FIG NO.	(b)	(2) SMR CODE a ARMY	b AIR FORCE	d NAVY	e USMC	(3) NATIONAL STOCK NUMBER	(4) DESCRIPTION REF NUMBER MFR CODE	USABLE ON CODE	(5) U/M	(6) QTY INC IN	(7) USMC QTY PER EQUIP
							GROUP 09 SKID BASE				
47	1	PAOZZ	PAOZZ	PAOZZ	PAOZZ	5310-00-696-5173	NUT, PLAIN, ASSEMBLED 501-250800-00 (78189) 69-561-5 (30554)	ABC	EA	2	2
47	2	PAOZZ	PAOZZ	PAOZZ	PAOZZ	5305-00-036-6906	SCREW, ASSEMBLED WASHER P-15121-50 (45722) 69-662-50 (30554)	ABC	EA 2	2	
47	3	PAOZZ	PAOZZ	PAOZZ	PAOZZ		CLAMP, LOOP MS21333-75 (96906)	ABC	EA	2	2
47	4	XBOZZ	XBOZZ	XBOZZ	XBOZZ	5340-00-050-2740	BRACKET, HOSE 319-0192 (44940) 84-13279 (30554)	ABC	EA 1	1	
47	5	PAOZZ	PAOZZ	PAOZZ	PAOZZ		NUT, PLAIN, HEXAGON MS51967-6 (96906)	ABC			
47	6	PAOZZ	PAOZZ	PAOZZ	PAOZZ	5310-00-931-8167	WASHER, LOCK MS35338-45 (9606) 50-1045 (44940)	ABC	EA 4	4	
47	7	PAOZZ	PAOZZ	PAOZZ	PAOZZ	5310-00-407-9566	WASHER, FLAT MS27183-12 (96906)	ABC	EA 4	4	
47	8	PAOZZ	PAOZZ	PAOZZ	PAOZZ	5310-00-481-4219	BOLT, MACHINE MS9072-32 (96906)	ABC	EA	4	4
47	9	PAOZZ	PAOZZ	PAOZZ	PAOZZ	5306-00-2264825	MOUNT, RESILIENT 25012-1 (81860) 4024572 (44940) 84-13034 (30554)	ABC	EA 8	8	
47	10	PAOZZ	PAOZZ	PAOZZ	PAOZZ	5340-01-277-3363	NUT, PLAIN, HEXAGON MS35649-225N (96906)	ABC	EA 4	4	
47	11	PAOZZ	PAOZZ	PAOZZ	PAOZZ		WASHER, LOCK MS35335-89 (96906)	ABC			
47	12	PAOZZ	PAOZZ	PAOZZ	PAOZZ	5310-01-301-1982	TERMINAL, STUD S-38230-G4 (74159) 13208E5820-3 (97403)	ABC	EA 1	1	
47	13	PAOZZ	PAOZZ	PAOZZ	PAOZZ	5310-00-942-5109	SCREW, DRIVE MS21318-14 (96906)	ABC	EA 1	1	
47	14	MDFZZ	MDOZZ	MDOZZ	MDOZZ		PLATE, IDENTIFICATION 1321 E6730 (49403)	ABC	EA 1	1	
47	15	MDOZZ	MDOZZ	MDOZZ	MDOZZ	5940-00-952-2327	PLATE, IDENT AUX FUE 75-5074 (30554)	ABC	EA	6	6
						5305-00-175-3230			EA	1	1
									EA 1	1	

(1) ILLUSTRATION (a) FIG NO.	(1) ILLUSTRATION (b)	(2) SMR CODE a ARMY	(2) SMR CODE b AIR FORCE	(2) SMR CODE d NAVY	(2) SMR CODE e USMC	(3) NATIONAL STOCK NUMBER	(4) DESCRIPTION REF NUMBER MFR CODE / USABLE ON CODE	(5) U/M	(6) QTY INC IN	(7) USMC QTY PER EQUIP
							GROUP 09 SKID BASE			
47	16	PAOZZ	PAOZZ	PAOZZ	PAOZZ	4730-00-909-8627	CLAMP, HOSE .... ABC MS35342-13 (96906)	EA	2	2
47	17	PAOOO	PAOOO	PAOOO	PAOOO	4720-00-021-3320	HOSE ASSY........ ABC 69-668 (30554) 501-115 (44940)	EA	1	1
47	18	XBFZZ	XBFZZ	XBFZZ	XBFZZ		SKID ASSY, ENG/GEN .... ABC 403-2361 (4940) 84-13013 (30554)	EA	1	1
47	19	PAOZZ	PAOZZ	PAOZZ	PAOZZ	5305-01478-5064	SCREW, ASSEMBLED........ ABC P-15121-78 (45722) 69-662-78 (30554)	EA	10	10
47	20	PAOZZ	PAOZZ	PAOZZ	PAOZZ		WASHER, FLAT .... ABC MS27183-10 (96906)	EA	10	10
47	2	PAOZZ	PAOZZ	PAOZZ	PAOZZ	5310408	BRACKET, MOUNTING........ ABC 403-2658 (44940) 84-13221 (30554)	EA	2	2
47	22	PAOZZ	PAOZZ	PAOZZ	PAOZZ	5340-01-275-3477	NUT, PLAIN, ASSEMBLED .... ABC 511-101800-00 (78189) 69-561-4 (30554)	EA	2	2
						5310-00-094-3421				

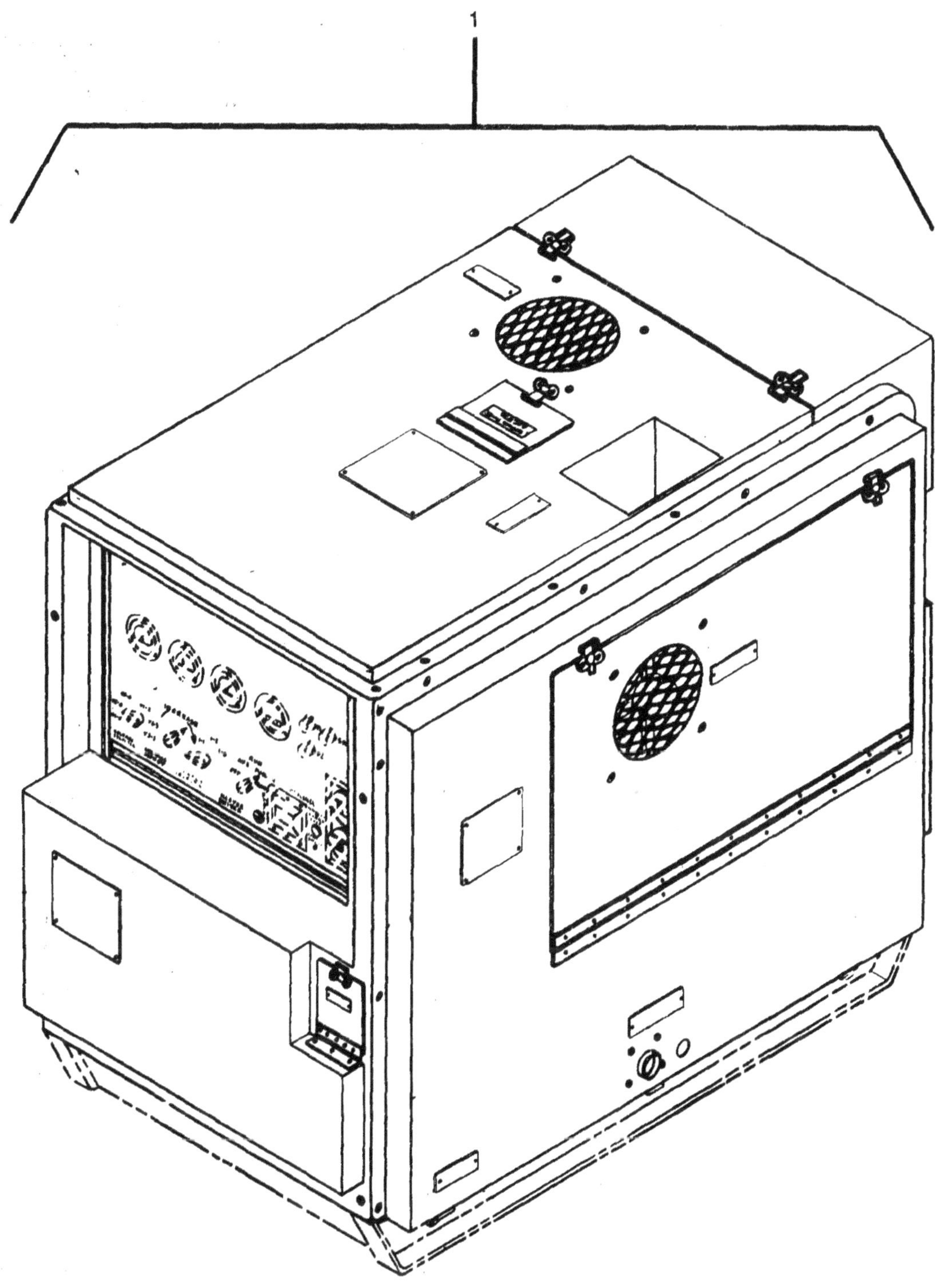

Figure 48. Acoustic Suppression Kit, Group 10.

(1) ILLUSTRATION (a) FIG NO.	(b)	(2) SMR CODE a ARMY	b AIR FORCE	d NAVY	e USMC	(3) NATIONAL STOCK NUMBER	(4) DESCRIPTION USABLE ON CODE REF NUMBER MFR CODE	(5) U/M	(6) QTY INC IN	(7) USMC QTY PER EQUIP
48	1	PAFZZ	PAFZZ	PAFZZ	PAFZZ	5180-01-279-9331	PULLER KIT, UNIVERSAL .... ABC EA 1 420-0516 (44940)			1
48	2	PAFZZ	PAFZZ	PAFZZ	PAFZZ	5120-01-279-4797	SOCKET, WRENCH.............................. ABC EA 1 420-0517 (44940)			1
48	3	PAFZZ	PAFZZ	PAFZZ	PAFZZ	5120-01-279-1658	TIMING TOOL .... ABC EA 1 420-0518 (44940)			1

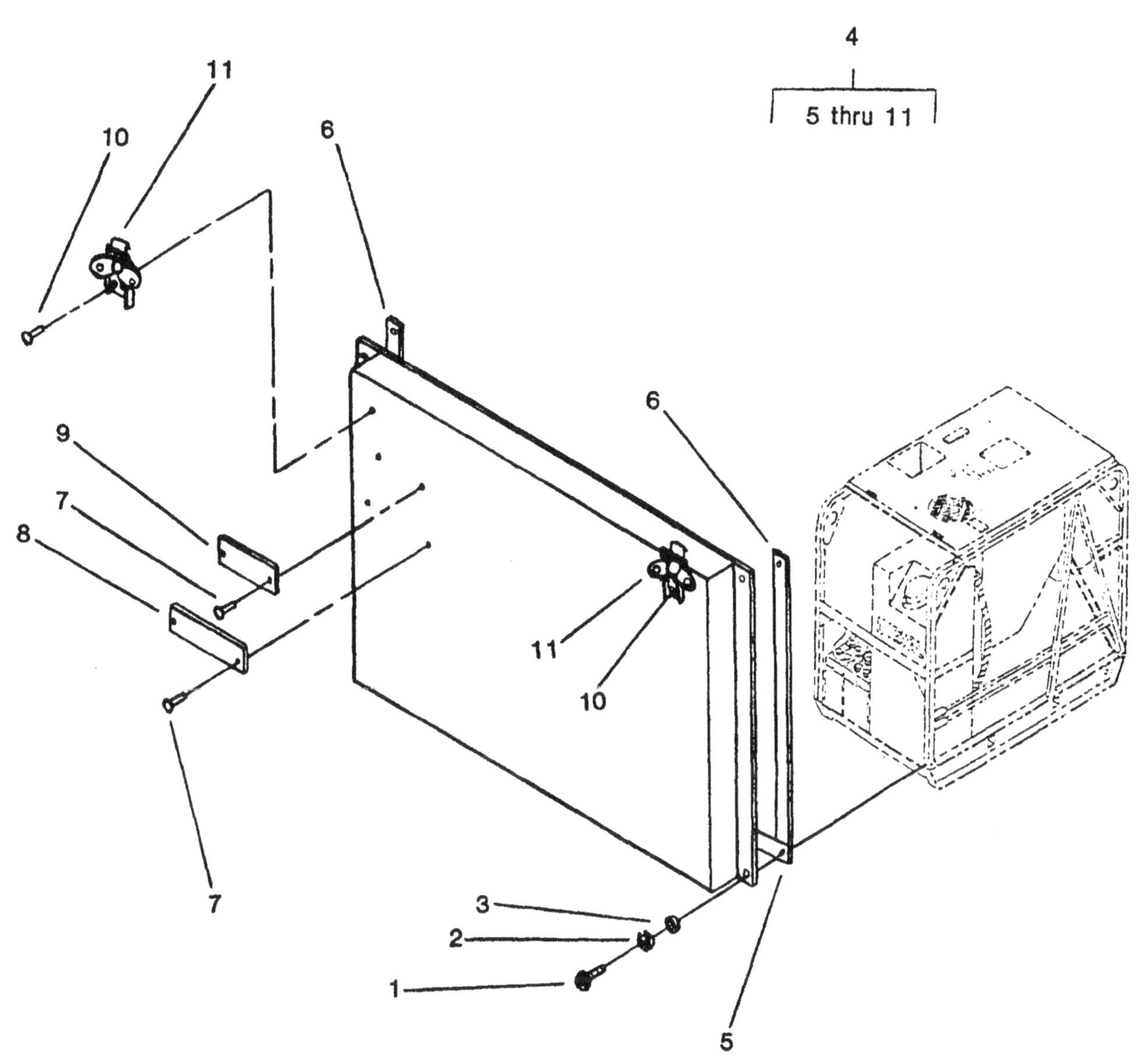

Figure 49. Panels, Group 10 (Sheet 1 of 7).

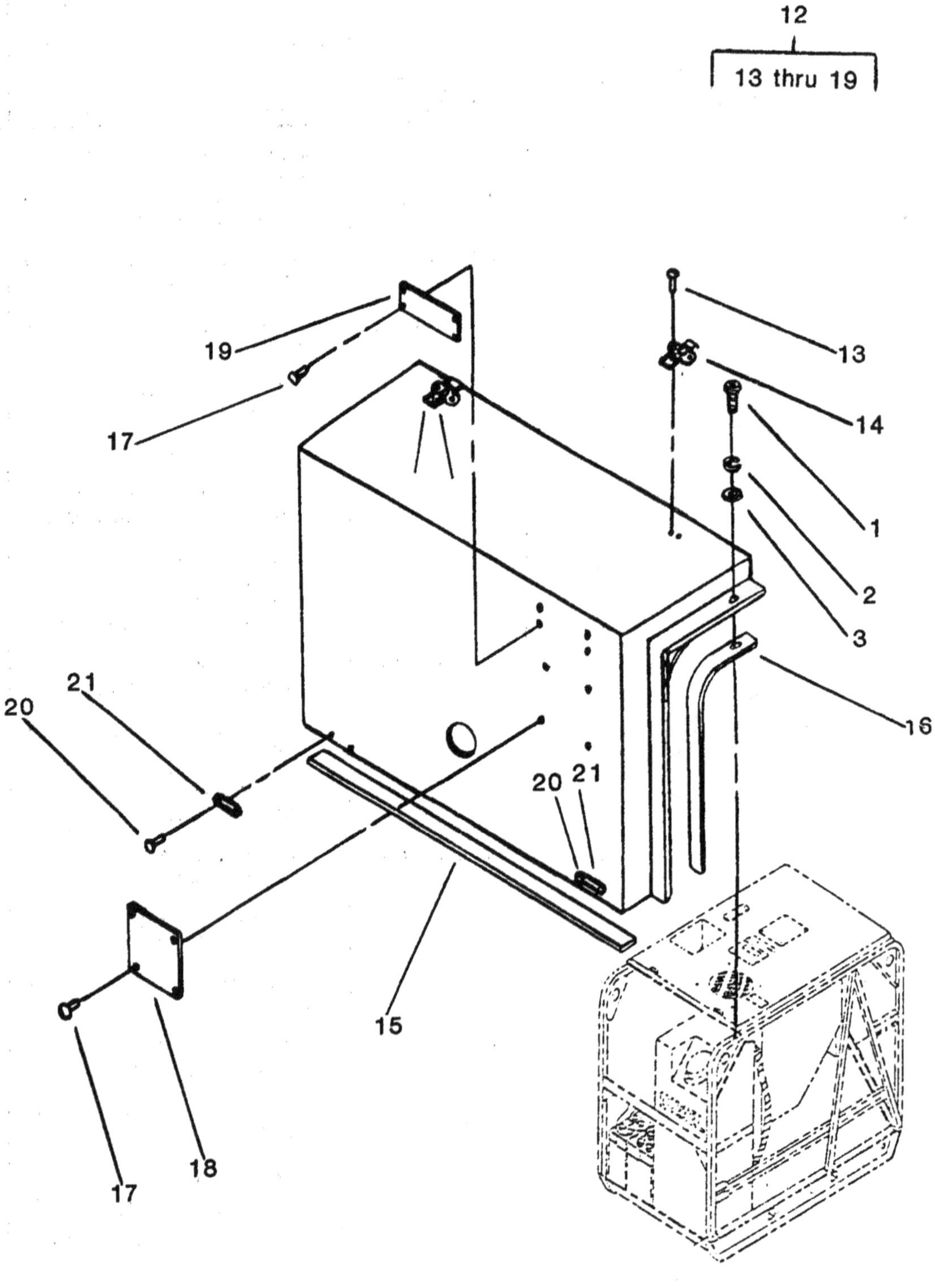

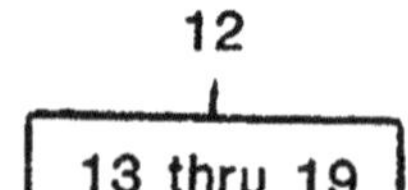

Figure 49. Panels, Group 10 (Sheet 2 of 7).

(1) (2) (3) (4) (5)(6)(7)
ILLUSTRA- SMR CODE
TION QTY USMC
ABABDE NATIONAL DESCRIPTIION USABLE INC QTY
FIG ITEM AIR STOCK ON IN PER
NO NO ARMY FORCE NAVY USMC NUMBER REF NUMBER MFR CODE CODE U/M UNITS EQUIP

GROUP 10 ACOUSTIC SUPPRESSION
KIT

49 1 PAOZZ 5305-00-059-3662 SCREW,PHN,10-32 A EA 42 50
MS51958-66 (96906)

49 2 PAOZZ 5310-00-933-8120 WASHER, LOCK, NO. 10 A EA 42 42
MS35338-138 (96906)

49 3 PAOZZ 5310-00-595-6772 WASHER, FLAT, NO. 10 A EA 42 50
MS15795-808 (96906)

49 4 MDOOO COVER ASSY ENG BOT A EA 1 1
88-13565 (30554)

49 5 MOOOO GASKET,12X1.00
X 21.5 IN. A EA 1
GRADE M BLACK (81349)
MAKE FROM 9320-00-855-3399

49 6 MOOOO GASKET, 12X1.00
X 13 IN. A EA 2 1
GRADE M BLACK (81349)
MAKE FROM 9320-00-855-3399

49 7 PAOZZ 5320-00-957-3582 RIVET,BLIND,CE A EA 4 38
M24243/6-A403H (81349)

49 8 MDOZZ PLATE,INST A EA 1 1
84-13014 (30554)

49 9 MDOZZ PLATE,IDENT A EA 1 1
88-13547-3 (30554)

49 10 PAOZZ 5320-00-956-7355 RIVET,BLIND CE A EA 8 28
M2424316-A604H (81349)

49 11 XBOZZ CATCH AND KEEPER A EA 2 4
8203-02 (03007)

49 12 MDOOO COVER ASSY ENG TOP A EA 1 1
88-13564 (30554)

49 13 PAOZZ 5320-00-956-7355 RIVET,BLIND CE A EA 8 28
M24243/6-A604H (81349)

49 14 XBOZZ CATCH AND KEEPER A EA 2 4
8203-02 (03007)

49 15 MOOOO GASKET, 12X1.00
X 21.5 IN. A EA 1 1
GRADE M BLACK (81349)
MAKE FROM 9320-00-855-3399

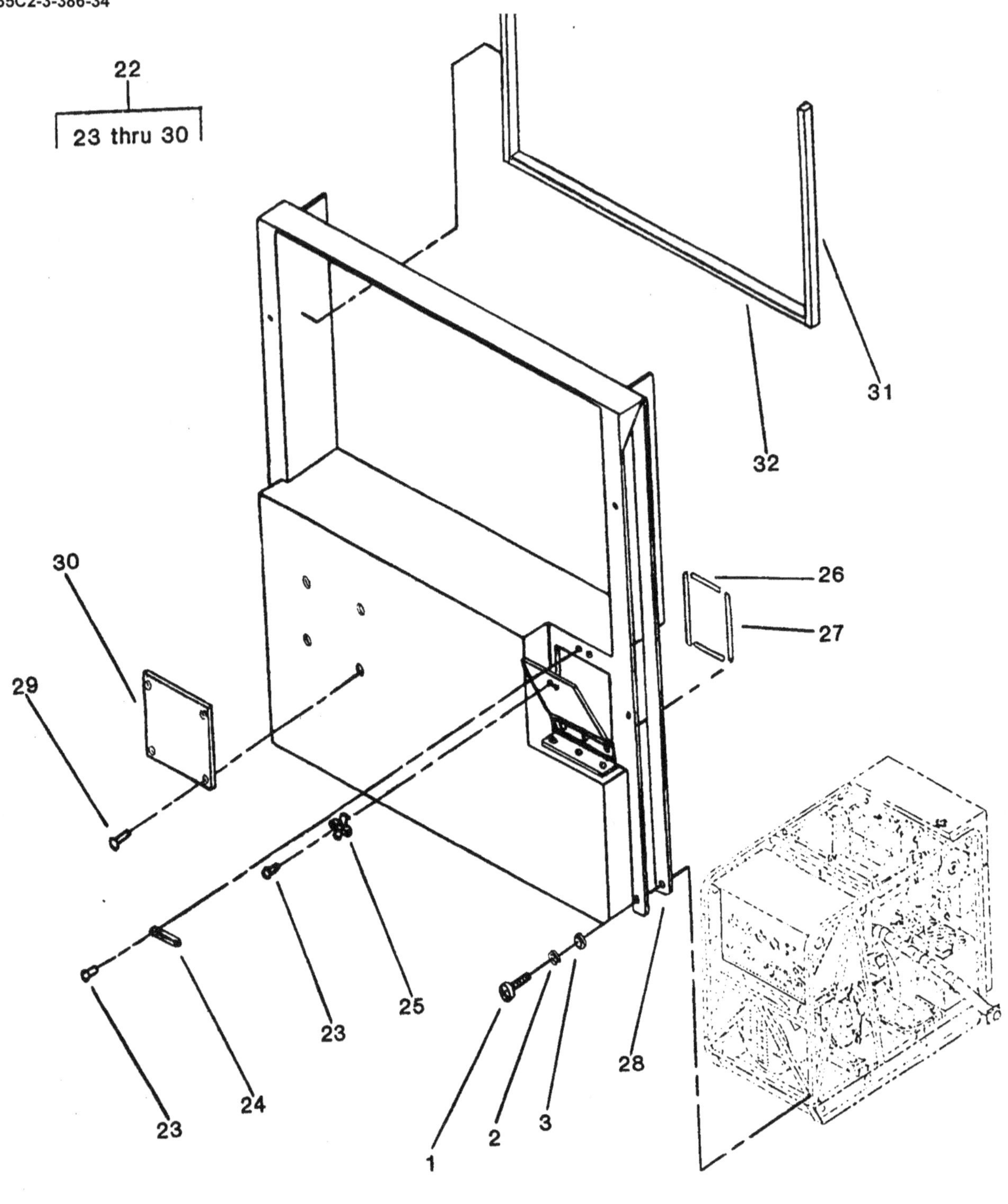

Figure 49. Panels, Group 10 (Sheet 3 of 7).

(1) (2) (3) (4) (5)(6)(7)
ILLUSTRA- SMR CODE
TION QTY USMC
ABABDE NATIONAL
FIG ITEM AIR STOCK DESCRIPTIION USABLE INC QTY
NO NO ARMY FORCE NAVY USMC NUMBER REF NUMBER MFR CODE CODE U/M UNITS EQUIP ON IN PER

GROUP 10 ACOUSTIC SUPPRESSION
KIT

49 16 MOOOO GASKET, 12X1.00
X 16.5 IN. A EA 2 2
GRADE M BLACK (81349)
MAKE FROM 9320-00-855-3399

49 17 PAOZZ 5320-00-957-3582 RIVET,BLIND,CE A EA 8 38
M24243/6-A403H (81349)

49 18 MOOZZ PLATE IDENT A EA 1 1
84-13038 (30554)

49 19 MOOZZ PLATE,IDENT A EA 1 1
88-13547-5 (30554)

49 20 PAOZZ 5320-00-956-7355 RIVET,BLIND CE A EA 6 28
M24243/6-A403H (81349

49 21 XBOZZ CATCH AND KEEPER A EA 2 4
8208-02 (03007)

49 22 MDOOO COVER ASSY INST A EA 1 1
88-13566 (30554)

49 23 PAOZZ 5320-00-952-4162 RIVET,BLIND CE A EA 4 16
AD43S (07707)

49 24 XBOZZ CATCH AND KEEPER A EA 1 4
8207-02 (03007)

49 25 XBOZZ CATCH AND KEEPER A EA 1 4
8201-02 (03007)

49 26 MOOOO RUBBER, U CHNL (4.5 IN.) A EA 2 2
X-172 (88786)
MAKE FROM 9390-00-641-3010

49 27 MOOOO RUBBER, U CHNL (5.0 IN.) A EA 2 2
X-172 (88786)
MAKE FROM 9390-00-641-3010

49 28 MOOOO GASKET, 12X1.00
X 25 INCHES A EA 2 2
GRADE M BLACK (81349)
MAKE FROM 9320-00-855-3399

49 29 PAOZZ 5320-00-957-3582 RIVET,BLIND,CE A EA 4 38
M24243/6-A403H (81349)

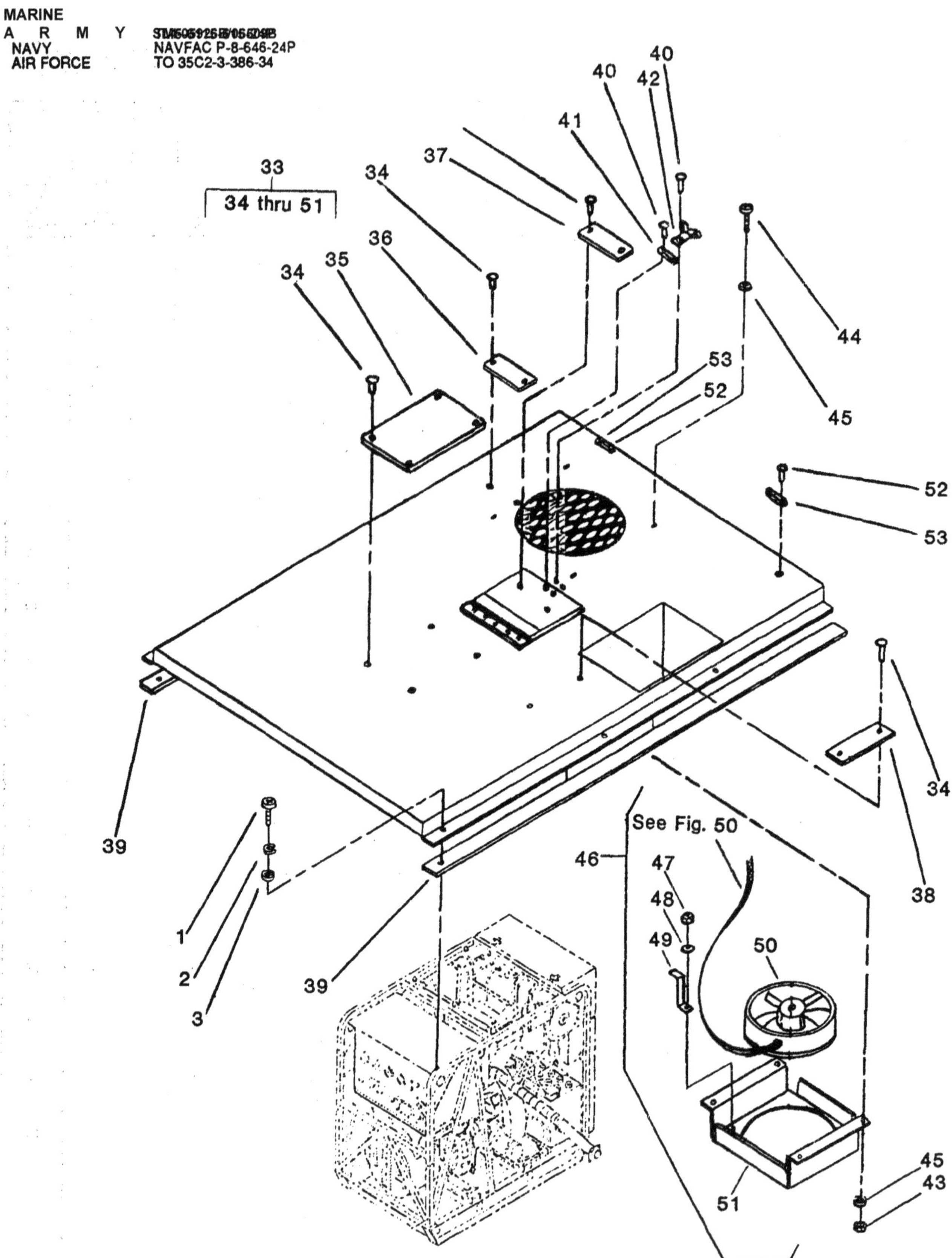

Figure 49. Panels, Group 10 (Sheet 4 of 7).

(1) (2) (3) (4) (5) (6) (7)
ILLUSTRA- SMR CODE
TION QTY USMC
ABABDE NATIONAL
FIG ITEM AIR STOCK DESCRIPTIION USABLE INC QTY
NO NO ARMY FORCE NAVY USMC NUMBER REF NUMBER MFR CODE CODE U/M UNITS EQUIP ON IN PER

GROUP 10 ACOUSTIC SUPPRESSION KIT

49 30 MDOZZ PLATE, OPER.INS A EA 1 1
88-13555 (30554)

49 31 MOOOO GASKET, .5X1.00
X 10 IN A EA 2 2
GRADE M BLACK (81349)
MAKE FROM 9320-00-X88-3202

49 32 MOOOO GASKET, .5X1.00
X 20 IN. A
GRADE M BLACK (81349)
MAKE FROM 9320-00-X88-3202

49 33 MDOOO COVER,ASSY,TOP A EA 1 1
88-13563 (30554)

49 34 PAOZZ 5320-00-957-3582 RIVET,BLIND,CE A EA 10 38
M2424316-A402H (81349)

49 35 PAOZZ PLATE,IDENT A EA 1 1
88-13603 (30554)

49 36 MDOZZ PLATE,CARTON A EA 1 1
88-13616 (30554)

49 37 MDOZZ PLATE,IDENT A EA 1 1
88-13547-2 (30554)

49 38 MDOZZ PLATE,CAUTION A EA 1 1
88-13548 (30554)

49 39 MOOOO GASKET, .12X1.00
X INCHES A EA 2 2
GRADE M BLACK (81349)
MAKE FROM 9320-00-855-3399

49 40 PAOZZ 5320-00-952-4162 RIVET,BLIND,CE A EA 4 16
M24243/6-A403H (81349)

49 41 XBOZZ CATCH AND KEEPER A EA 1 4
8207-02 (03007)

49 42 XBOZZ CATCH AND KEEPER .25 A EA 1 4
8201-02 (03007)

49 43 XBOZZ 5310-00-877-5797 NUT,SELF-LOCK A EA 4 16
MS21044-N3 (96906)

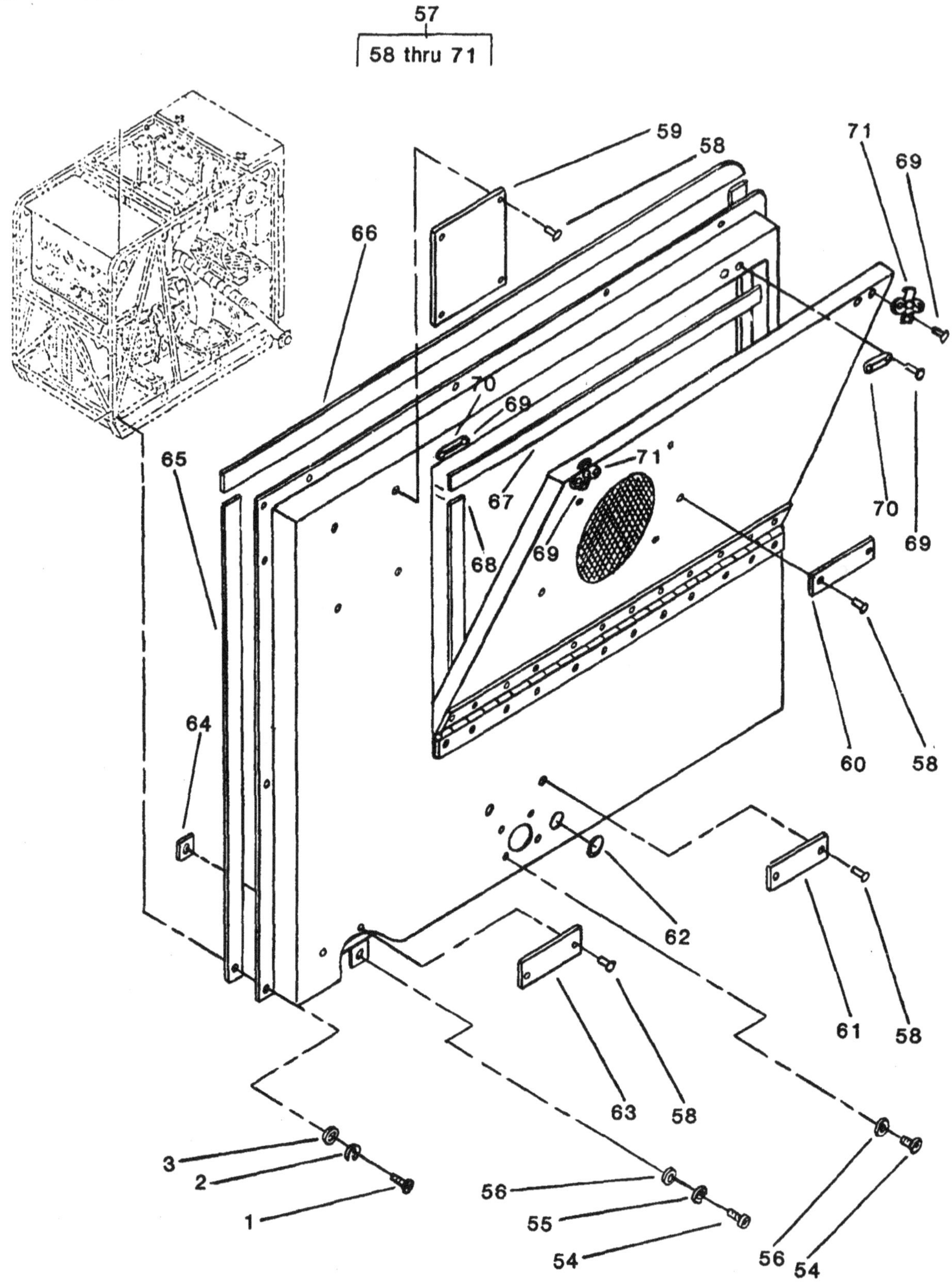

Figure 49, Panels, (3roup 10 (Sheet 5 of 7).

(1) ILLUSTRATION FIG NO	(2) ILLUSTRATION ITEM NO	(3) SMR CODE ARMY / AIR FORCE / NAVY / USMC (QTY USMC ABABDE)	(4) NATIONAL STOCK NUMBER	(5)(6)(7) DESCRIPTIION	REF NUMBER (MFR CODE)	USABLE ON CODE	U/M	QTY INC IN UNITS	QTY PER EQUIP
				GROUP 10 ACOUSTIC SUPPRESSION KIT					
49	44	PAOZZ	5305-00-059-3662	SCREW,PNH,10-32	MS51958-66 (9690/)	A	EA	4	50
49	45	PAOZZ	5310-00-595-6772	WASHER,FLAT,NO. 10	MS15795-808 (96906)	A	EA	4	50
49	46	MDOOO		FAN ASSY	88-13598-002 (30554)	A	EA	1	1
49	47	XBOZZ	5310-00-877-5797	NUT,SELF-LOCK	MS21044-C08 (96906)	A	EA	4	16
49	48	PAOZZ	5310-00-225-5328	WASHER,FLAT	MS15795-841 (96906)	A	EA	4	8
49	50	PAOZZ	4140-00-925-5188	FAN, TUBE AXL	M23071/1-001 (81349)	A	EA	1	1
49	51	XBOZZ		BRACKET, FAN MT	88-13596-001 (30554)	A	EA	1	1
49	52	PAOZZ	5320-00-956-7355	RIVET,BLIND,CE	M24243/6-A403H (81349)	A	EA	6	28
49	53	XBOZZ		CATCH AND KEEPER	8208-02 (03007)	A	EA	2	4
49	54	PAOZZ	5305-00-071-1318	SCREW, .25-20 UNC	MS51957-83 (96906)	A	EA	10	10
49	55	PAOZZ	5310-00-933-8121	WASHER,LOCK, NO. (.25)	MS35338-139 (96906)	A	EA	6	6
49	56	PAOZZ	5310-00-582-5677	WASHER,FLAT,NO. (.25)	MS15795-810 (96906)	A	EA	10	10
49	57	MDOOO		COVER,ASSY RS	88-13561 (30554)	A	EA	1	1
49	58	PAOZZ	5310-00-957-3582	RIVET,BLIND,CE	AD42S (07707)	A	EA	10	38
49	59	MDOZZ		PLATE,INST	88-13558 (30554)	A	EA	1	1
49	60	MDOZZ		PLATE,IDENT	88-13547-6 (30554)	A	EA	1	1

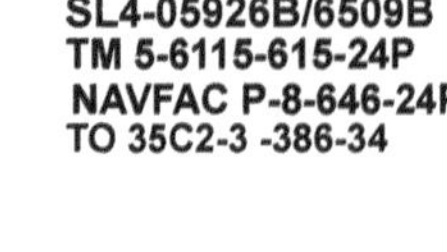

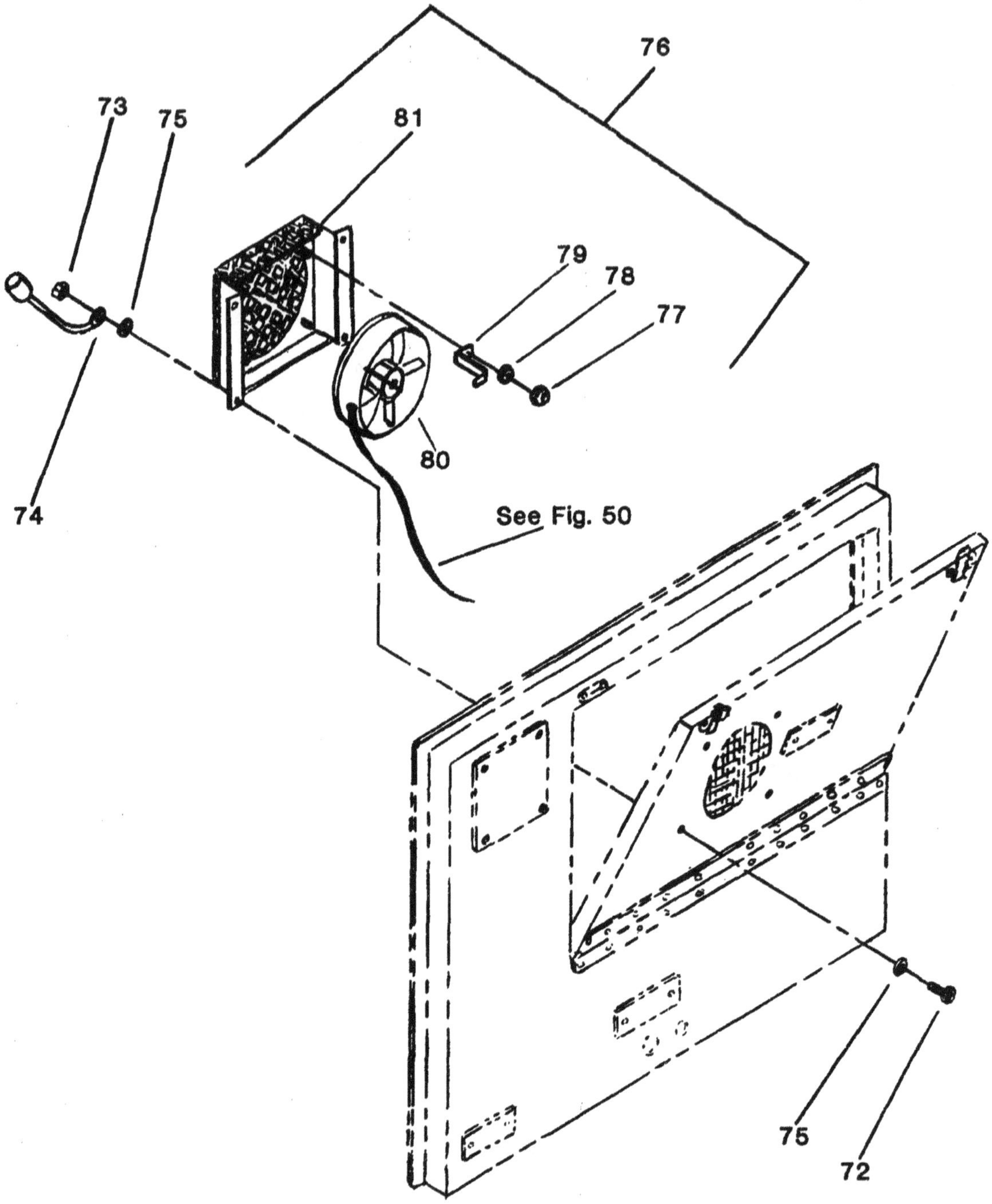

Figure 49. Panels, Group 10 (Sheet 6 of 7].

(1) (2) (3) (4) (5) (6) (7)
ILLUSTRA- SMR CODE
TION QTY USMC
ABABDE NATIONAL DESCRIPTIION USABLE INC QTY
FIG ITEM AIR STOCK ON IN PER
NO NO ARMY FORCE NAVY USMC NUMBER REF NUMBER MFR CODE CODE U/M UNITS EQUIP

GROUP 10 ACOUSTIC SUPPRESSION KIT

49 61 MDOZZ PLATE,DANGER A EA 1 1
84-13549 (30554)

49 62 XBOZZ GROMMET,RUBBER A EA 1 1
AN931A16-22 (88044)

49 63 MDOZZ PLATE,IDENT A EA 1 1
88-13546-4 (30554)

49 64 MOOOO GASKET,(.12X1.0X13.25) A EA 3 6
GRADE M BLACK (81349)
MAKE FROM 9320-00-855-3399

49 65 MOOOO GASKET, (.12X1.0) A EA 2 4
GRADE M BLACK (81349)
MAKE FROM 9320-00-855-3399

49 66 MOOOO GASKET, (.12X1.0) A EA 4 4
GRADE M BLACK (81349)
MAKE FROM 9320-00-855-3399

49 67 MOOOO GASKET, (.12X1.0X39) A EA 1 1
GRADE M BLACK (81349)
MAKE FROM 9320-00-855-3399

49 68 MOOOO GASKET, (.12X1.0X25) A EA 2 2
GRADE M BLACK
MAKE FROM 9320-00-855-3399

49 69 PAOZZ 5320-00-952-4162 RIVET,BLIND CE A EA 10 33
M2424 3/6-A403H (81349)

49 70 XBOZZ CATCH AND KEEPER A EA 2 4
8207-02 (03007)

49 71 XBOZZ CATCH AND KEEPER A EA 2 4
8201-02 (03007)

49 72 PAOZZ 5305-00-059-3662 SCREW,PNH,10-32 A EA 4 50
915025-0066 (65597)

49 73 XBOZZ 5310-00-877-5797 NUT,SELF-LOCK A EA 4 16
88-13561 (30554)

49 74 PAOZZ WIRE ASSY,BONDING A EA 1 1
88-13587-002 (30554)
WASHER,FLAT,NO. 10 A
49 75 PAOZZ 5310-00-595-6772 MS15795-808 (96906) EA 8 50

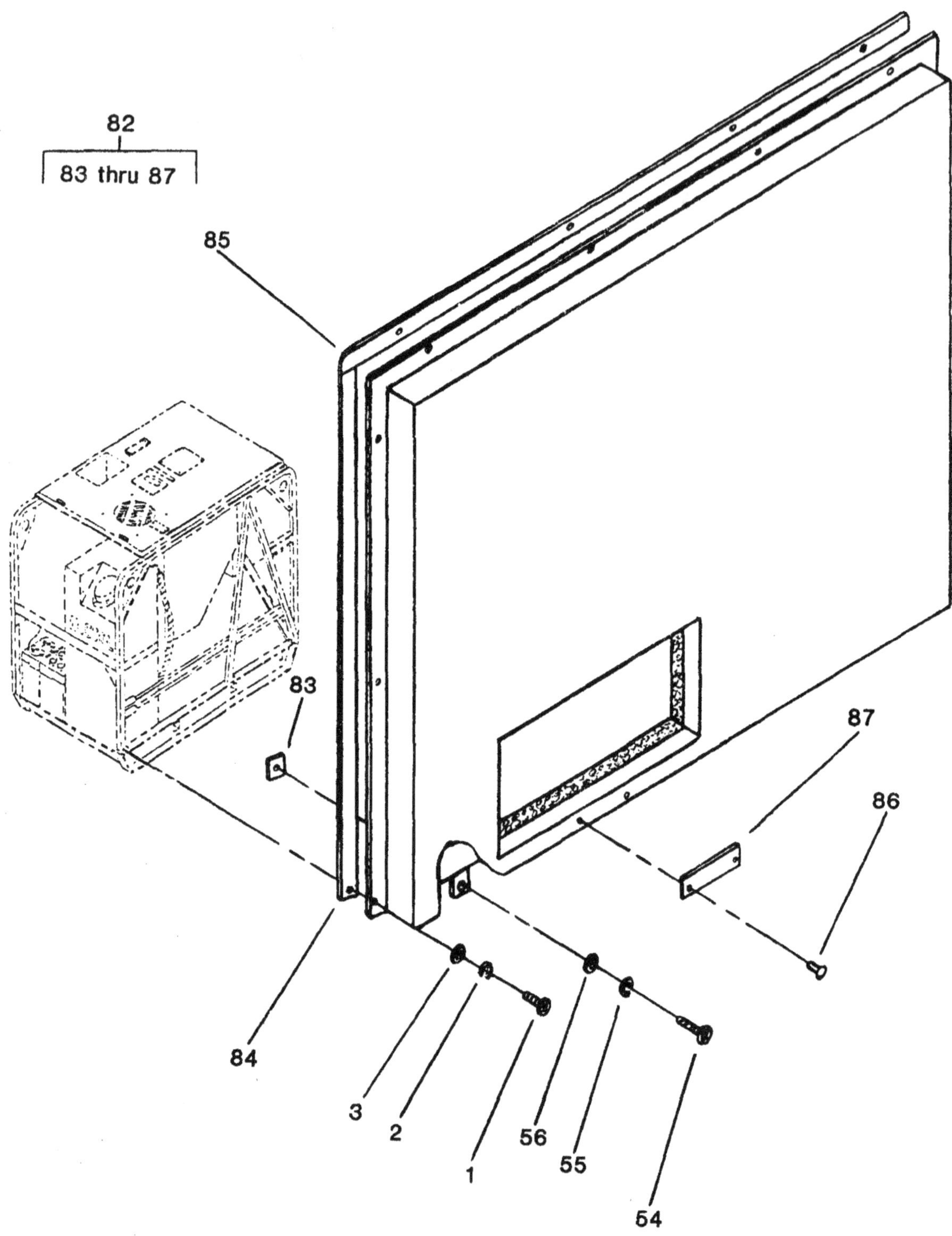

Figure 49. Panels, Group 10 (Sheet 7 of 7).

(1) (2) (3) (4) (5)(6)(7)
ILLUSTRA- SMR CODE
TION QTY USMC
ABABDE NATIONAL DESCRIPTIION USABLE INC QTY
FIG ITEM AIR STOCK ON IN PER
NO NO ARMY FORCE NAVY USMC NUMBER REF NUMBER MFR CODE CODE U/M UNITS EQUIP

GROUP 10 ACOUSTIC SUPPRESSION
KIT

49 76 MOOOO FAN ASSY A EA 1 1
88-13598-001 (30554)

49 77 XBOZZ 5310-00-877-5797 NUT,SELF-LOCK A EA 4 16
MS21044-C08 (96906)

49 78 PAOZZ 5310-00-225-5328 WASHER,FLAT A EA 4 8
MS15795-841 (96906)

49 79 XBOZZ CLIP,FAN MT A EA 8 8
88-13597 (30554)

49 80 PAOZZ 4140-00-925-5188 FAN,TUBE,AXL A EA 1 2
M23071/1-001 (81349)

49 81 MDOZZ BRKT ASSY FAN MT A EA 1 1
88-13596-001 (30554)

49 82 MOOOO COVER ASSY LS A EA 1 1
88-13562 (30554)

49 83 MOOOO GASKET, (.12X1.0X13.25) A EA 3 6
GRADE M BLACK
MAKE FROM 9320-00-855-3399

49 84 MOOOO GASKET,(.12X1.0) A EA 2 2
GRADE M BLACK (81349)
MAKE FROM 9320-00-855-3399

49 85 MOOOO GASKET, (.12X1.0X24.5) A EA 1 1
GRADE M BLACK (81349)
MAKE FROM 9320-00-855-3399

49 86 PAOZZ 5320-00-957-3582 RIVET,BLIND,CE A EA 2 38
M24243/6-A402H (81349)

49 87 MDOZZ PLATE,IDENT A EA 1 1
88-13546-5 (30554)

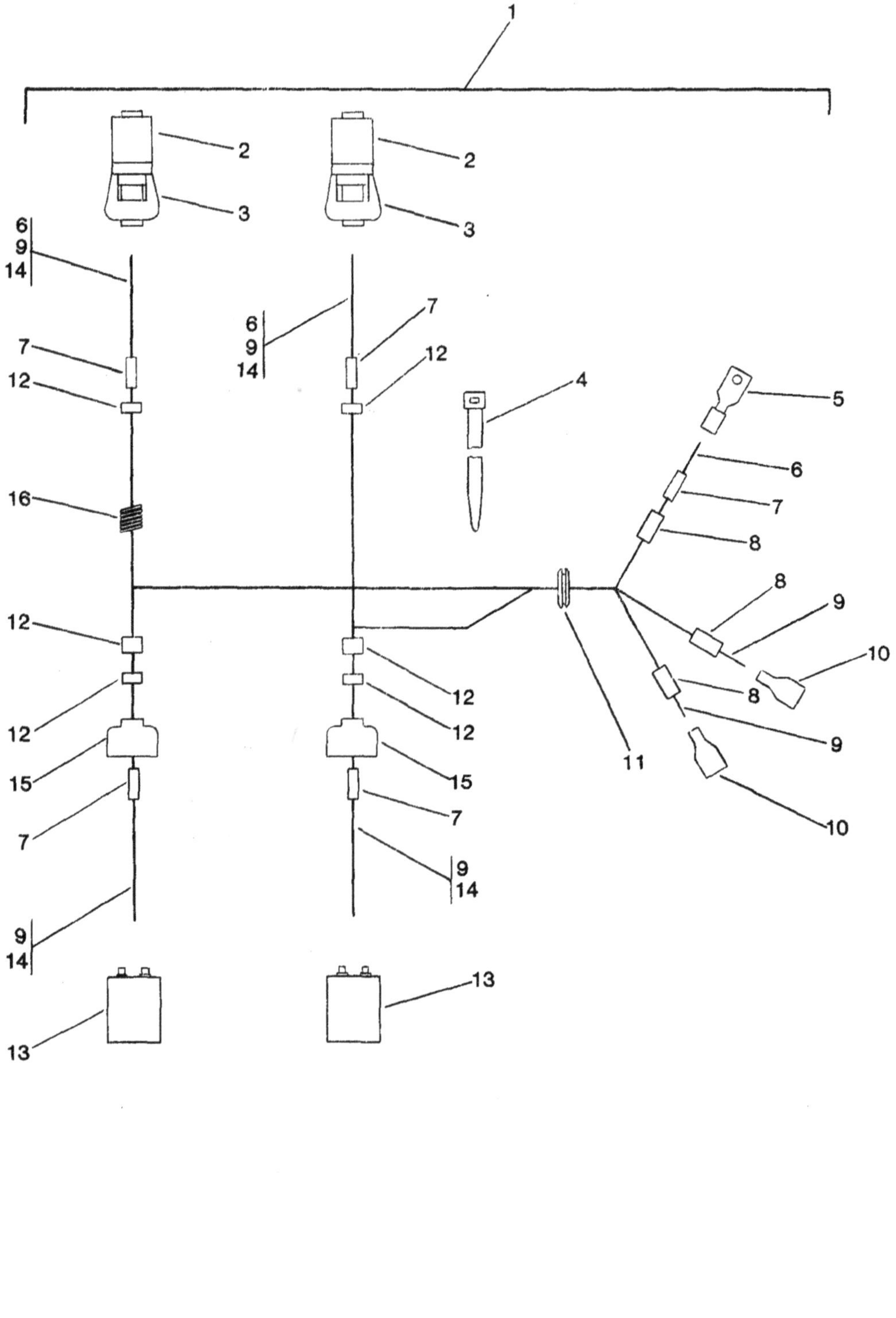

Figure 50. Dual Fan Wire Harness, Group 10 (Acoustic Suppression Kit).

(1) ILLUSTRATION FIG NO	(2) ITEM NO	(3) SMR CODE ARMY	AIR FORCE	NAVY	USMC	(4) NATIONAL STOCK NUMBER	(5) DESCRIPTION REF NUMBER (MFR CODE)	(6) USABLE ON CODE	(7) U/M	INC IN UNITS	QTY PER EQUIP
							GROUP 10 ACOUSTIC SUPPRESSION KIT				
50	1	MDOOO					WIRE HARN DUAL BLO 88-13605 (30554)	A	EA	1	1
50	2	XADZZ				5935-00-815-1541	CONNECTOR MS3476W8-4S (96906)	A	EA	2	2
50	3	ZADZZ					ADAPTER,CONNECTOR M8504952-1-8W (81349)	A	EA	2	2
50	4	PAOZZ				5975-00-074-2072	STRAP,TIE DOWN MS3367-1-9 (96906)	A	EA	AR	AR
50	5	PAOZZ				5940-00-825-3699	LUG,TERMINAL MS17143-10 (96906)	A	EA	1	1
50	6	XADZZ					WIRE,ELEC 20 AWG M16878/4GBG9 (81349)	A	EA	AR	AR
50	7	MFOZZ				5970-00-954-1622	INSULATION SHRINK FIT2213/16 BALCK (92194) MAKE FROM SHRINK P/N M23053/ 5-105-0	A	FT	AR	AR
50	8	MFOZZ				5970-00-787-2325	INSULATION SHRINK FIT2211-8 YELLOW (92194) MAKE FROM SHRINK P/N M23053/ 5-104-4	A	FT	AR	AR
50	9	XADZZ					WIRE,ELEC 20 AWG M16878/4BGB0 (81349)	A	FT	AR	AR
50	10	PAOZZ				5940-01-112-9746	RCPT,FEMALE 2-520184-2 (00779)	A	EA	2	2
50	11	XAOZZ					GROMMET Z2258-9 (76385)	A	EA	1	1
50	12	MFOZZ				5970-00-787-2321	INSULATION,SHRINK FIT2213-32 YEELOW (92194) MAKE FROM SHRINK P/N M23053/ 5-103-4	A	FT	AR	AR
50	13	PAOZZ					CAPACITOR KGDX2020 (14655)	A	EA	2	2
50	14	XADZZ					WIRE,ELEC 20 AWG M16878/4BGB2 (81349)	A	EA	2	2

(1) ILLUSTRATION A FIG NO	B ITEM NO	(2) QTY	(3) SMR CODE A ARMY	B AIR FORCE	D NAVY	E USMC	(4) NATIONAL STOCK NUMBER	(5) (6) (7) DESCRIPTIION REF NUMBER MFR CODE	USABLE ON CODE	U/M	INC QTY IN UNITS	QTY PER EQUIP
								GROUP 10 ACOUSTIC SUPPRESSION KIT				
50	15							BOOT, TERMINAL 30397 (14655)	A	EA	2	2
50	16							TUBING, SPIRAL WRAP SW-2 (92194) MAKE FROM TUBING P/N SW-2	A	FT	AR	AR

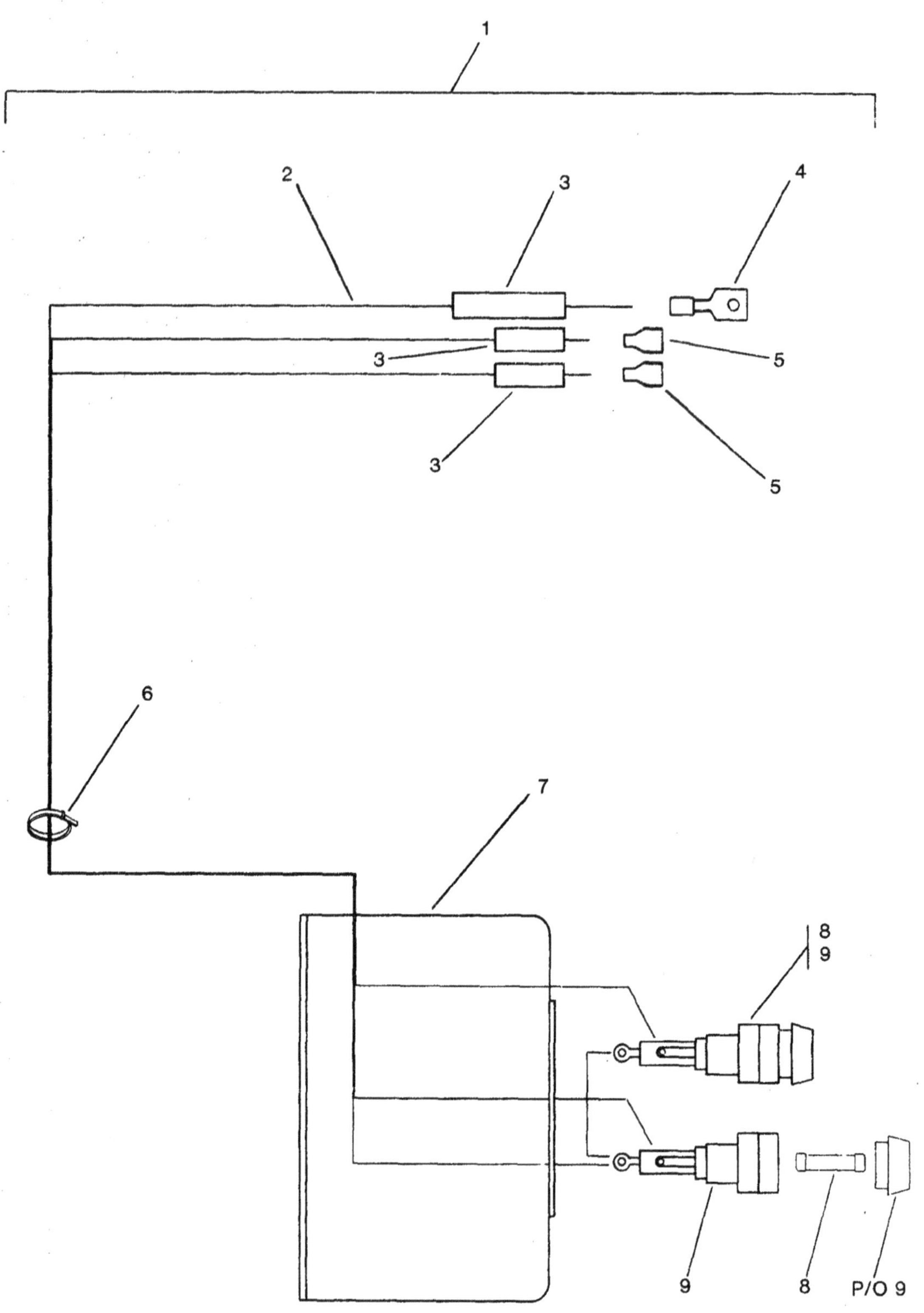

Figure 51. Fuse Block Assembly, Group 10 (Acoustic Suppression Kit).

(1) ILLUSTRATION FIG NO	(2) ILLUSTRATION ITEM NO	(3) SMR CODE ABABDE ARMY AIR FORCE NAVY USMC	(4) NATIONAL STOCK NUMBER	(5)(6)(7) DESCRIPTIION REF NUMBER MFR CODE	USABLE ON CODE	U/M	INC IN UNITS	QTY PER EQUIP
				GROUP 10 ACOUSTIC SUPPRESSION KIT				
51	1	MDOOO		FUSE BLOCK ASSY 88-13540 (30554)	A	EA	1	1
51	2	XADZZ		WIRE,ELEC 20 AWG M16878/4BGB0 (81349)	A	FT	AR	AR
51	3	MFOZZ	5970-00-787-2325	INSULATION,SHRINK FIT2211-8 YELLOW (92194) MAKE FROM SHRINK P/N M23053/ 5-104-4	A	FT	AR	AR
51	4	PAOZZ	5940-00-825-3699	LUG,TERMINAL MS17143-10 (96906)	A	EA	1	1
51	5	PAOZZ	5940-01-126-3973	TAB,INSULATED MALE RA25177 (59730)	A	EA	2	2
51	6	PAOZZ	5975-00-074-2072	STRAP,TIE DOWN MS3367-1-9 (96906)	A	EA	AR	AR
51	7	XBOZZ		BRACKET,FUSE HOLD 88-13529 (30554)	A	FT	1	1
51	8	PAOZZ	5920-00-284-9220	FUSE,FAN ASSY FO2B250V1A 2110-0312 (28480)	A	EA	2	2
51	9	PAOZZ	5920-00-892-9311	HOLDER,FUSEA FHN26G1 (81349)		EA	2	2

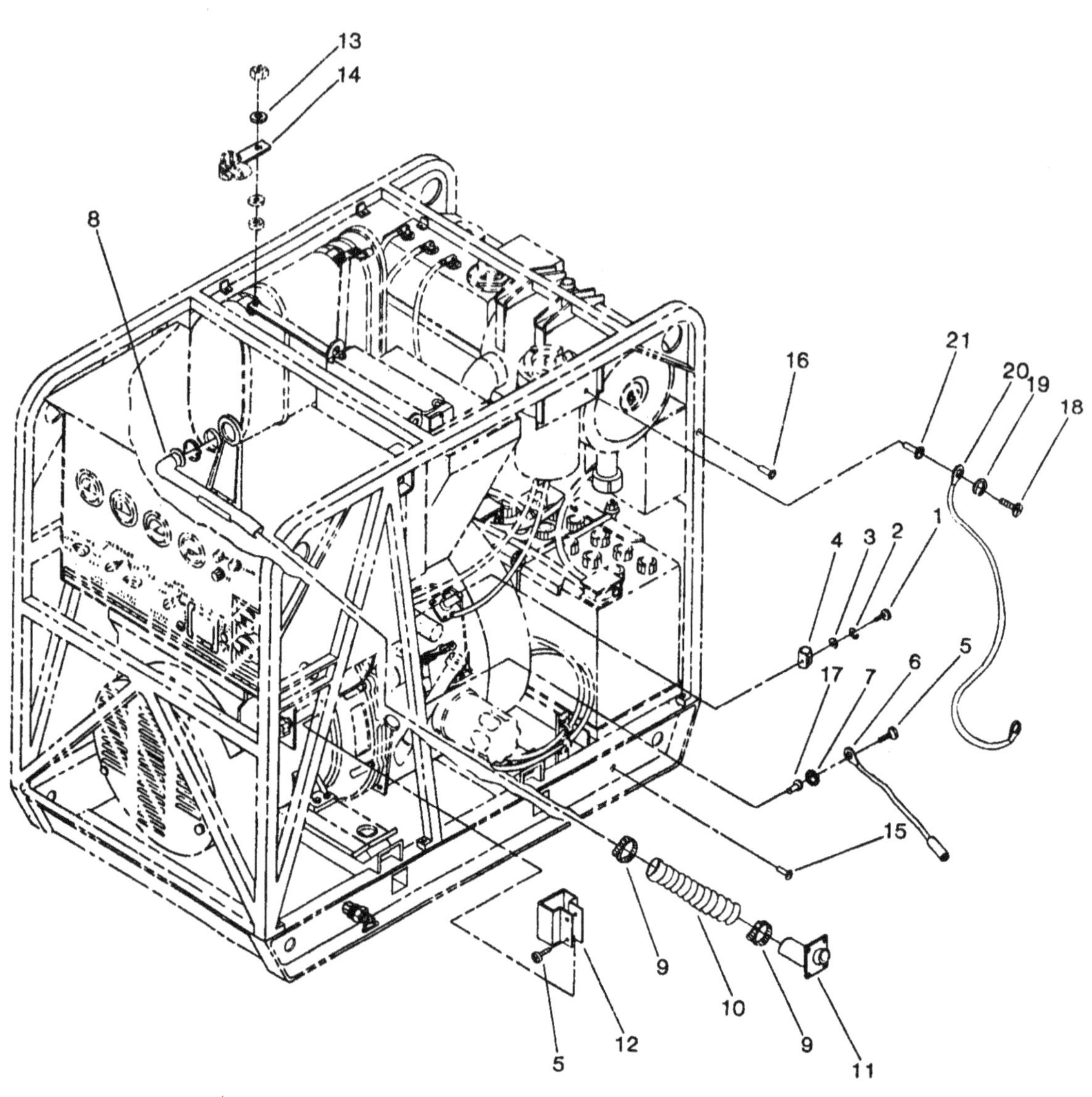

Figure 52. Modified Generator ASK Items, Group 10 (Acoustic Suppression Kit).

(1) (2) (3) (4) (5)(6)(7)
ILLUSTRA- SMR CODE
TION QTY USMC
ABABDE NATIONAL DESCRIPTIION USABLE INC QTY
FIG ITEM AIR STOCK ON IN PER
NO NO ARMY FORCE NAVY USMC NUMBER REF NUMBER MFR CODE CODE U/M UNITS EQUIP

GROUP 10 ACOUSTIC SUPPRESSION KIT

FIG NO	ITEM NO	SMR CODE	NATIONAL STOCK NUMBER	DESCRIPTION / REF NUMBER (MFR CODE)	USABLE ON CODE	U/M	QTY INC IN UNITS	QTY PER EQUIP
52	1	PAOZZ	5305-00-059-3639	SCREW,PNH,10-32 MS51958-63 (96906)	A	EA	2	2
52	2	PAOZZ	5310-00-933-8120	WASHER,LOCK NO. 10 MS35338-138 (96906)	A	EA	2	2
52	3	PAOZZ	5310-00-595-6772	WASHER,FLAT NO. 10 MS15795-808 (96906)	A	EA	2	2
52	4	PAOZZ	5340-00-682-1617	CLAMP,LOOP TYPE MS21919DG12 (96906)	A	EA	2	2
52	5	PAOZZ	5305-00-432-4163	SCREW,TAPPING MS51861-24 (96906)	A	EA	8	8
52	6	PAOZZ		WIRE,BONDING 88-13587-001 (30554)	A	EA	1	1
52	7	PAOZZ	5310-00-209-1239	WASHER,LOCK MS35335-60 (96906)	A	EA	1	1
52	8	MDOZZ		TUBE,MOD EXH 88-13613 (30554)	A	EA	1	1
52	9	PAOZZ	4730-00-908-3193	CLAMP,HOSE MS35842-12 (96906)	A	EA	2	2
52	10	XBOZZ		TUBE,FLEXIBLE ONA-1005 (16632)	A	EA	1	1
52	11	XBOZZ		PIPE,EXH,WLD ASSY 88-13612 (30554)	A	EA	1	1
52	12	XBOZZ		BRACKET,CAPACITOR 88-1350-1 (30554)	A	EA	2	2
52	13	PAOZZ	5310-00-974-6623	WASHER,LOCK 4917260-05 (90536)	A	EA	2	2
52	14	XBOZZ		CLAMP,CABLE EXTD 88-13617 (30554)	A	EA	2	2
52	15	PAOZZ	5310-01-108-8404	NUT,PLAIN,BLD RVT MS27130-534 (96906)	A	EA	6	6
52	16	PAOZZ	5310-00-440-9231	NUT,PLAIN,BLD RVT NAS132903-80 (80205)	A	EA	44	44

(1) ILLUSTRATION FIG NO	(2) ITEM NO	(3) SMR CODE ARMY / AIR FORCE / NAVY / USMC	(4) NATIONAL STOCK NUMBER	(5) (6) (7) DESCRIPTIION REF NUMBER MFR CODE	USABLE ON CODE	U/M	QTY INC IN UNITS	USMC QTY PER EQUIP
				GROUP 10 ACOUSTIC SUPPRESSION KIT				
52	17	PAOZZ	5310-01-046-5371	NUT,PLAIN,BLD RVT MS27130-S25 (96906)	A	EA	2	2
52	18	PAOZZ	5305-00-059-3662	SCREW,PNH,10-32 MS51958-66 (96906)	A	EA	1	REF
52	19	PAOZZ	5310-00-933-8120	WASHER,LOCK NO. 10 MS35338-138 (96906)	A	EA	1	1
52	20	XBOZZ		CABLE ASSY,LANYARD 52305A18-6 (84256)	A	EA	1	1
52	21	PAOZZ	5310-00-990-1361	NUT,PLAIN,BLD RIVET MS27130-525 (96906)	A	EA	1	2

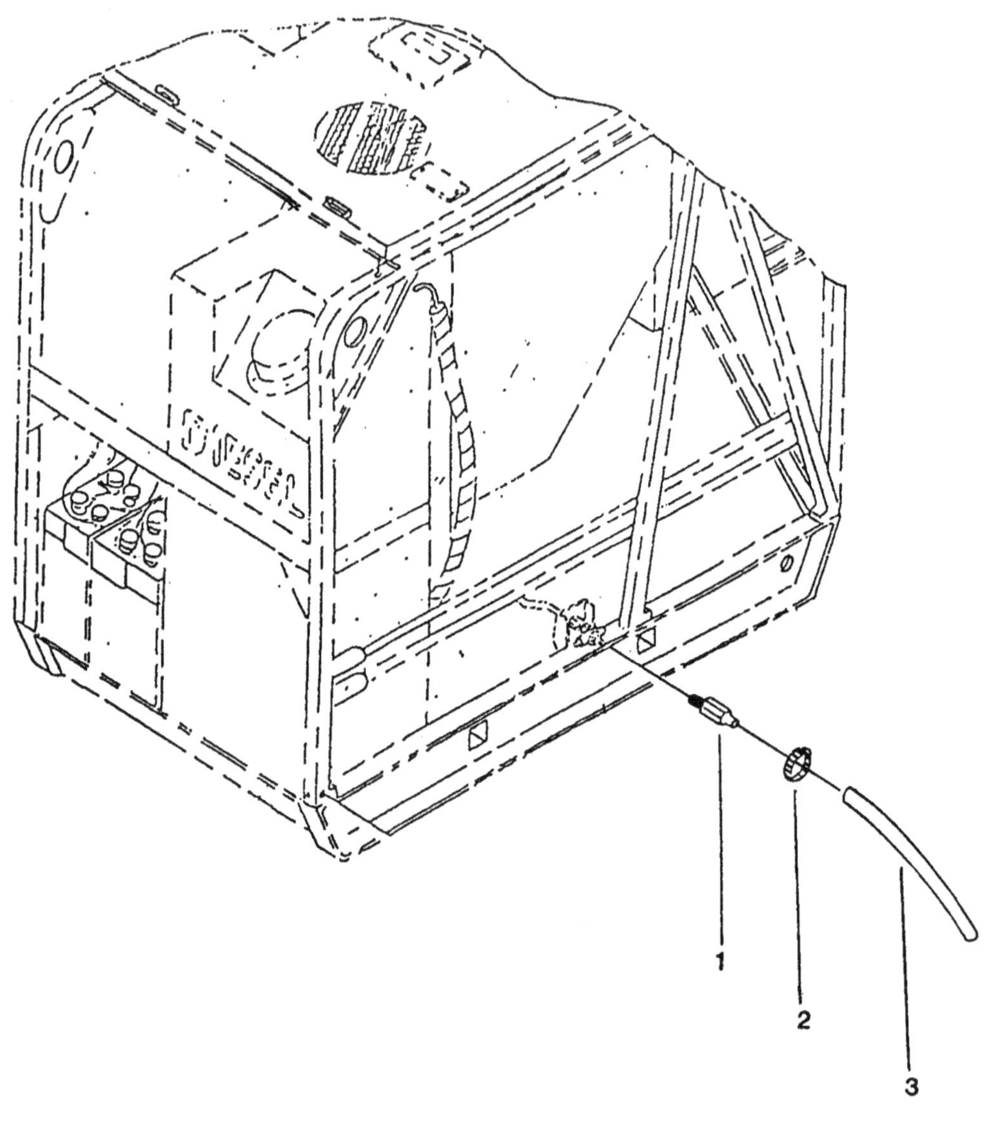

Figure 53. Oil Drain Adapter, Hose, and Clamp, Group 10 (Acoustic Suppression Kit).

(1) (2) (3) (4) (5) (6) (7)
ILLUSTRA- SMR CODE
TION QTY USMC
ABABDE NATIONAL
FIG ITEM AIR STOCK DESCRIPTIION USABLE INC QTY
NO NO ARMY FORCE NAVY USMC NUMBER REF NUMBER MFR CODE CODE U/M UNITS EQUIP ON IN PER

GROUP 10 ACOUSTIC SUPPRESSION
KIT

53 1 XBOZZ ADAPER,OIL DRAIN A EA 1 1
88-13544 (30554)

53 2 XBOZZ 4730-00-908-3194 CLAMP,HOSE A EA 1 1
MS35842-11 (96906)

53 3 XBOZZ HOSE,RUBBER A EA 1 1
M6000-E-00108 (96906)
MAKE FROM RUBBER HOSE
P/N MIL-H-6000.625 ID X 20
INCHES LONG

(1) (2) (3) (4) (5) (6) (7)
ILLUSTRA- SMR CODE
TION QTY USMC
ABABDE NATIONAL DESCRIPTIION USABLE INC QTY
FIG ITEM AIR STOCK ON IN PER
NO NO ARMY FORCE NAVY USMC NUMBER REF NUMBER MFR CODE CODE U/M UNITS EQUIP

GROUP II: BULK MATERIALS
FIG. BULK

BULK PAOZZ 9320-00-855-3399 TAPE,SPONGE RUBBER FT V
SELF-ADHESIVE, 1/8 IN. THICK
GRADE M BLACK (81349)

BULK PAOZZ 9320-00-X88-3202 TAPE,SPONGE RUBBER FT V
SELF-ADHESIVE, 1/2 IN. THICK,
GRADE M BLACK (81349)

BULK PAOZZ 9390-00-641-3010 RUBER,U CHNL FT V
X-172 (88786)

BULK PAOZZ SHRINK,INSULATION FT V
M23053/5-103-4 (92194)

BULK PAOZZ SHRINK,INSULATION FT V
M23053/5-104-4 (92194)

BULK PAOZZ SHRINK,INSULTION FT V
M23053/5-105-0 (92194)

BULK PAOZZ TUBING,SPIRAL WRAP FT V
SW-2 (92194)

Section III. SPECIAL TOOIS LIST

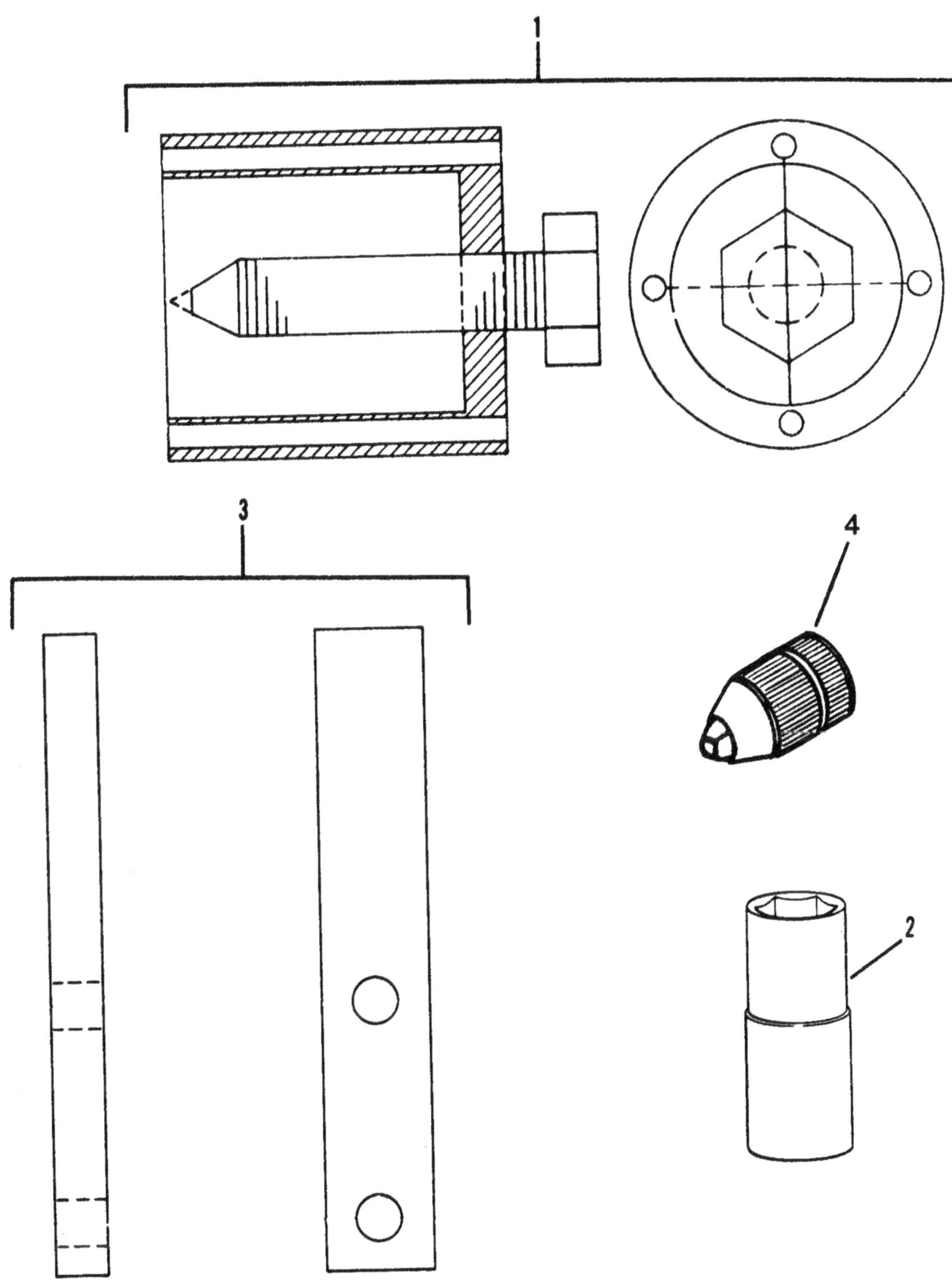

Figure 54. Special Tools.

(1) ILLUSTRATION FIG NO	(2) ITEM NO	(3) SMR CODE ABABDE ARMY AIR FORCE NAVY USMC	(4) NATIONAL STOCK NUMBER	(5) (6) (7) DESCRIPTIION REF NUMBER MFR CODE	USABLE ON CODE	U/M	INC IN UNITS	QTY PER EQUIP
54	1	PAFZZ		PULLER,CRANK GEAR 420-0516 (44940)	ABC	EA	1	1
54	2	PAFZZ		SOCKET,ROCKER ARM 420-0517 (44940)	ABC	EA	1	1
54	3	PAFZZ		BAR,TIMING FIXTURE 420-0515 (44940)	ABC	EA	1	1
54	4	PAOZZ		STOP, DRILL BIT U1577 (90788)	A	EA	1	1

Section IV. NATIONAL STOCK NUMBER AND REFERENCE NUMBER INDEX

Table 3. National Stock Number to Figure and Item Number Index - Continued

NSN	FIGURE NO..	ITEM NO.	NSN	FIGURE NO.	ITEM NO.
1560-01-280-7086	31	4	2910-01-276-1484	7	1
2510-00-534-5828	36	10	2920-01-223-8762	17	6
2510-01-275-5349	13	12	2920-01-224-6246	17	7
2520-01-223-8780	17	9	2920-01-224-6247	17	5
2520-01-275-1732	8	12	2920-01-274-6728	17	4
2590-01-274-9241	24	4	2920-01-274-6763	17	10
2590-01-274-9241	31		2920-01-274-9479	27	2
2810-01-276-1027	25	11	2920-01-275-1621	17	1
2815-01-212-4008	17	14	2920-01-275-1694	17	18
2815-01-274-6764	26	6	2920-01-275-4311	21	8
2815-01-274-6768	23	4	2920-01-276-6903	32	27
2815-01-274-6803	12	1	2920-01-282-8522	35	22
2815-01-274-6813	28	7	2920-01-282-8522	43	59
2815-01-274-6814	24	7	2930-01-275-1639	13	11
2815-01-274-8213	24	8	2930-01-275-4249	13	14
2815-01-274-8214	24	9	2930-01-276-5970	31	25
2815-01-274-9239	28		2940-01-274-6800	9	6
2815-01-274-9240	28		2940-01-274-8456	22	6
2815-01-274-9254	24	10	2940-01-274-8456	31	
2815-01-274-9255	24	11	2940-01-275-4285	31	22
2815-01-274-9354	31	31	2940-01-275-9157	31	23
2815-01-274-9361	31	28	2990-01-274-9376	21	19
2815-01-274-9368	24	6	2990-01-275-1717	15	12
2815-01-274-9369	21	21	2990-01-275-1757	15	15
2815-01-274-9386	23	5	3020-01-225-6988	17	11
2815-01-274-9474	21	3	3020-01-274-9402	29	5
2815-01-274-9514	28		3020-01-274-9403	29	6
2815-01-274-9516	23	8	3020-01-274-9404	30	11
2815-01-275-0178	23	7	3040-01-049-0578	20	19
2815-01-275-1707	29	9	3040-01-274-6809	11	3
2815-01-275-1711	30	12	3040-01-275-0203	20	18
2815-01-275-1712	24	1	3040-01-275-0205	20	21
2815-01-275-2484	23	3	3040-01-275-1640	20	14
2815-01-275-4331	24	1A	3040-01-275-2537	18	10
2815-01-275-5364	28		3040-01-287-9617	29	7
2815-01-275-5368	24	2	3110-00-158-8243	32	29
2815-01-276-3515	31	5	3110-00-158-8243	33	29
2815-01-294-1816	6	25	3110-01-218-0623	17	15
2910-00-893-6402	7	5	3110-01-236-6408	17	16
2910-00-911-0081	6	27	3120-01-276-3375	30	5
2910-01-063-3144	7	2	3120-01-276-3375	31	20
2910-01-274-6761	21	14	3120-01-276-3385	28	11
2910-01-274-6767	18	14	3120-01-276-3389	30	10
2910-01-274-9375	21	13	3120-01-276-8639	26	2
2910-01-275-1715	20	27	3120-01-277-2403	28	12
2910-01-275-1749	6	31	3120-01-277-4435	31	9
2910-01-275-2460	10	15	3120-01-277-8898	31	18
2910-01-275-2460	10		3120-01-280-6902	13	7
2910-01-275-8028	10	6	3120-01-281-9234	31	13
2910-01-275-9144	18		4010-00-809-2719	6	9
2910-01-275-9144	18		4020-00-523-9641	37	67
2910-01-276-1483	18	8	4020-00-523-9641	43	22

Section IV. NATIONAL STOCK NUMBER AND REFERENCE NUMBER INDEX

Table 3. National Stock Number to Figure and Item Number Index - Continued

NSN	FIGURE NO..	ITEM NO. NSN	FIGURE NO. ITEM NO.		
4030-00-270-5436	6	8	4820-00-136-1085	6	32
4030-00780-9350	37	65	4820-00-595-3191	6 19	
4030-00-780-9350	43	20	4820-01-274-6684 10	8	
4140-00-925-5188	49	50	4820-01-282-0165	9	10
4140-00-925-5188	49	80	5120-01-278-4797 48	2	
4140-01-274-7721	26	3	5120-01-279-1658 48	3	
4330-01-281-5988	9 11	5180-01-279-9331 48		1	
4710-01-274-4922	6 22	5305-00-036-6906		6	28
4710-01-274-8341	20	6	5305-00-036-6906 37 24		
4710-01-275-5381	18	4	5305-00-036-6906 43 34		
4720-00-021-3320	47	17	5305-00-036-6906 47	2	
4720-01-212-2604	18	3	5305-00-036-6968 37 52		
4720-01-274-4749	8	1	5305-00-036-6968 43 23		
4720-01-274-4749	8	2	5305-00-036-6970 37	6	
4720-01-274-4932	22	13	5305-00-036-6970 37 28		
4720-01-274-4933	9	5	5305-00-036-6970 43 26		
4720-01-275-2441	6 16	5305-00-036-6972 37		2	
4720-01-275-5225	6 36	5305-00-036-6972 40 17			
4720-01-275-6168	6	3	5305-00-036-6972 43 14		
4720-01-275-6169	6 17	5305-00-036-6976 37		4	
4720-01-281-5193	10	2	5305-00-036-6976 43 24		
4720-01-288-0786	10	1	5305-00-036-6977 37 55		
4730-00-011-2578	15	16	5305-00-036-6977 43 10		
4730-00-277-7939	6	4	5305-00-036-6978 35	9	
4730-00-278-4497	6 10	5305-00-036-6978 37 23			
4730-00-289-5484	18	5	5305-00-036-6978 43	2	
4730-00-483-5176	7	4	5305-00-038-3089 43 60		
4730-00-540-1861	6 23	5305-00-035-3103 37 22			
4730-00-570-2932	15	11	5305-00-038-3103 43 38		
4730-40-808-6814	30	14	5305-00-038-3122 36 12		
4730-00-808-6814	31	12 5305-00-038-3145 35 14			
4730-00-812-1333	6	7	5305-00-038-3145 37 56		
4730-00-824-0497	6	5	5305-00-038-3145 43 42		
4730-00-844-3308	31	26	5305-00-059-3639 52	1	
4730-00-871-6729	18	13	5305-00-059-3662 49	1	
4730-00-908-3193	52	9	5305-00-059-3662 49 44		
4730-00-908-3194	22	12	5305-00-059-3662 49 72		
4730-00-908-3194	53	2	5305-00-059-3662 52 18		
4730-00-908-6292	9	4	5305-00-068-0500 32	5	
4730-00-909-8627	47	16	5305-00-068-0500 33	5	
4730-01-050-3941	6 34	5305-00-068-0502 32 24			
4730-01-095-5584	7	6	5305-00-068-0502 33 24		
4730-01-103-1199	6	6	5305-00-071-1318 49 54		
4730-01-110-9055	18	1	5305-00-071-2510 32	2	
4730-01-274-6700	15	13	5305-00-071-2510 33	2	
4730-01-275-4175	20	8	5305-00-175-3230 32 13		
4730-01-275-4176	20	9	5305-00-175-3230 33 13		
4730-01-275-4180	31	32	5305-00-175-3230 47 13		
4730-01-276-9257	7 11	5305-00-177-5778 37 61			
4730-01-277-1388	6 20	5305-00-177-5778 43 17			
4730-01-277-1399	6 33	5305-00-191-6226		8	8
4730-01-287-8895	15	14	5305-00-211-9344 35 19		

Section IV. NATIONAL STOCK NUMBER AND REFERENCE NUMBER INDEX

Table 3. National Stock Number to Figure and Item Number Index - Continued

NSN	FIGURE NO..	ITEM NO.	NSN	FIGURE NO.	ITEM NO.
5305-00-211-9344	40	19	5306-00-226-4825	9	1
5305-00-211-9344	42	14	5306-00-226-4825	32	21
5305-00-224-1092	35	17	5306-00-226-4825	33	21
5305-00-224-1092	40	6	5306-00-226-4825	47	8
5305-00-224-1092	42	6	5306-00-226-4830	5	28
5305-00-224-1093	43	40	5306-01-174-8739	31	1
5305-00-269-3219	46	1	5306-01-228-7458	28	9
5305-00-432-4163	52	5	5306-01-238-3172	32	17
5305-00-776-9564	36	3	5306-01-275-1963	20	32
5305-00-776-9564	37	1	5306-01-275-3240	26	4
5305-00-776-9564	40	24	5306-01-275-3241	13	19
5305-00-776-9564	42	13	5306-01-275-3241	27	4
5305-00-776-9564	43	46	5306-01-275-3242	21	1
5305-00-889-3001	45	2	5306-01-275-3243	30	1
5305-00-942-2196	46	9	5306-01-275-3267	13	6
5305-00-958-5477	37	20	5306-01-275-5001	13	22
5305-00-958-5477	43	73	5306-01-275-5002	15	1
5305-00-984-6191	32	10	5306-01-275-5005	13	13
5305-00-984-6191	33	10	5306-01-275-6000	11	9
5305-01-078-5064	4	39	5306-01-275-6000	13	1
5305-01-078-5064	5	11	5306-01-275-6000	20	4
5305-01-078-5064	6	2	5306-01-275-6000	31	29
5305-01-078-5064	6	38	5306-01-275-6001	16	6
5305-01-078-5064	8	3	5306-01-275-6002	18	11
5305-01-078-5064	10	5	5306-01-276-0833	4	10
5305-01-078-5064	11	5	5306-01-276-7463	26	1
5305-01-078-5064	43	51	5306-01-276-7463	27	10
5305-01-078-5064	47	19	5306-01-277-3178	17	3
5305-01-114-5801	35	12	5306-01-277-3179	27	5
5305-01-114-5801	43	58	5306-01-278-1963	20	32
5305-01-147-8224	4	38	5306-01-280-6685	4	8
5305-01-203-5147	46	14	5307-01-275-3424	15	8
5305-01-212-3220	15	17	5307-01-275-3425	21	11
5305-01-212-3221	18	6	5307-01-275-3425	21	17
5305-01-212-3221	29	1	5307-01-276-7534	31	14
5305-01-212-3227	16	4	5307-01-277-1165	23	6
5305-01-212-3371	17	2	5310-00-022-8834	36	8
5305-01-226-6624	22	AI	5310-00-043-0520	20	31
5305-01-275-3287	25	5	5310-00-045-3299	32	11
5305-01-275-3291	20	37	5310-00-045-3299	33	11
5305-01-276-0850	13	8	5310-00-045-4007	45	3
5305-01-276-1627	13	16	5310-00-045-5210	36	9
5305-01-276-1628	22	3	5310-00-052-3632	35	13
5305-01-276-1629	10	13	5310-00-052-3632	36	11
5305-01-276-1637	20	26	5310-00-052-3632	37	19
5305-01-276-9193	29	4	5310-00-052-3632	43	11
5305-01-277-3184	20	22	5310-00-063-7360	37	48
5305-01-277-4990	20	1	5310-00-063-7360	40	16
5305-01-282-8181	27	1	5310-00-063-7360	43	13
5306-00-225-8496	4	23	5310-00-081-4219	4	19
5306-00-225-8496	37	43	5310-00-081-4219	5	19
5306-00-225-8496	43	32	5310-00-081-4219	5	27
5306-00-226-4825	5	17	5310-00-081-4219	6	13

Section IV. NATIONAL STOCK NUMBER AND REFERENCE NUMBER INDEX

Table 3. National Stock Number to Figure and Item Number Index - Continued

NSN	FIGURE NO..	ITEM NO.	NSN	FIGURE NO.	ITEM NO.
5310-00-081-4219	9	3	5310-00-696-5173	11	4
5310-00-081-4219	31	3	5310-00-696-5173	35	11
5310-00-081-4219	37	41	5310-00-696-5173	36	2
5310-00-081-4219	43	30	5310-00-696-5173	37	63
5310-00-081-4219	47	7	5310-00-696-5173	42	11
5310-00-094-3421	35	8	5310-00-696-5173	43	45
5310-00-094-3421	37	8	5310-00-696-5173	46	12
5310-00-094-3421	43	1	5310-00-696-5173	47	1
5310-00-094-3421	47	22	5310-00-732-0558	4	9
5310-00-167-0837	36	6	5310-00-732-0559	36	4
5310-00-184-8970	43	75	5310-00-761-0654	46	7
5310-00-186-9550	6	35	5310-00-761-0654	46	17
5310-00-187-2425	36	16	5310-00-761-6882	36	15
5310-00-187-2425	37	74	5310-00-809-4058	4	37
5310-00-187-2429	37	16	5310-00-809-4058	5	10
5310-00-199-1789	7	7	5310-00-809-4058	10	4
5310-00-209-0965	32	18	5310-00-809-4058	46	13
5310-00-209-0965	33	18	5310-00-809-4058	47	20
5310-00-209-1239	52	7	5310-00-809-4061	32	19
5310-00-225-5328	49	48	5310-00-809-4061	33	19
5310-00-225-5328	49	78	5310-00-836-3520	35	16
5310-00-229-4677	46	2	5310-00-836-3520	37	27
5310-00-229-4677	46	10	5310-00-836-3520	40	5
5310-00-407-9566	4	18	5310-00-836-3520	42	5
5310-00-407-9566	5	18	5310-00-836-3520	43	57
5310-00-407-9566	5	26	5310-00-869-1018	20	38
5310-00-407-9566	6	12	5310-00-877-5797	49	43
5310-00-407-9566	9	2	5310-00-877-5797	49	47
5310-00407-9566	15	3	5310-00-877-5797	49	73
5310-00-407-9566	31	2	5310-00-877-5797	49	77
5310-00-407-9566	32	22	5310-00-880-7744	4	11
5310-00-407-9566	33	22	5310-00-897-6082	37	15
5310-00-407-9566	37	40	5310-00-905-4600	6	11
5310-00-437-9566	43	29	5310-00-931-8167	4	17
5310-00-407-9566	47	6	5310-00-931-8167	5	25
5310-00-440-9231	52	16	5310-00-931-8167	37	39
5310-00-550-1130	32	25	5310-00-931-8167	43	28
5310-00-582-5677	49	56	5310-00-931-8167	47	5
5310-00-582-5965	32	3	5310-00-933-8120	49	2
5310-00-582-5965	33	3	5310-00-933-8120	52	2
5310-00-582-5965	33	25	5310-00-933-8120	52	19
5310-00-595-6772	49	3	5310-00-933-8121	49	55
5310-00-595-6772	49	45	5310-00-934-9751	20	20
5310-00-595-6772	49	75	5310-00-942-5109	47	11
5310-00-595-6772	52	3	5310-00-950-0440	37	75
5310-00-595-7237	36	5	5310-00-950-0440	43	76
5310-00-627-6128	46	8	5310-00-957-3582	49	58
5310-00-637-9541	46	18	5310-00-974-6623	52	13
5310-00-696-5173	4	36	5310-00-990-1361	52	21
5310-00-696-5173	5	9	5310-01-010-2261	46	21
5310-00-696-5173	6	1	5310-01-046-5371	52	17
5310-00-696-5173	10	3	5310-01-049-2745	20	12

Section IV. NATIONAL STOCK NUMBER AND REFERENCE NUMBER INDEX

Table 3. National Stock Number to Figure and Item Number Index - Continued

NSN	FIGURE NO..	ITEM NO.	NSN	FIGURE NO.	ITEM NO.
5310-01-049-4077	46	6	5315-00-449-2945	35	6
5310-01-108-8404	52	15	5315-00-449-2945	40	13
5310-01-137-4829	21	9	5315-00-449-2945	42	18
5310-01-212-3338	15	6	5315-00-652-8675	29	8
5310-01-212-4612	37	17	5315-00-652-8675	30	16
5310-01-227-6099	23	1	5315-00-839-5820	11	1
5310-01-240-1472	28	8	5315-01-275-3451	30	6
5310-01-275-3298	16	1	5315-01-275-3451	31	7
5310-01-275-3299	21	5	5315-01-275-6984	30	15
5310-01-275-3301	20	5	5315-01-276-9216	21	12
5310-01-275-3307	6	37	5320-00-165-8772	5	13
5310-01-275-3310	23	2	5320-00-165-8772	37	10
5310-01-275-3318	11	10	5320-00-165-8772	43	4
5310-01-275-3318	13	2	5320-00-395-6523	5	7
5310-01-275-3318	16	2	5320-00-395-6523	35	3
5310-01-275-3318	18	7	5320-00-395-6523	37	13
5310-01-275-3318	20	2	5320-00-395-6523	40	3
5310-01-275-3318	25	6	5320-00-395-6523	42	3
5310-01-275-3318	27	6	5320-00-395-6523	43	7
5310-01-275-3318	29	2	5320-00-952-4162	49	23
5310-01-275-3318	31	30	5320-00-952-4162	49	40
5310-01-275-3319	20	17	5320-00-952-4162	49	69
5310-01-275-3320	21	6	5320-00-956-7355	49	10
5310-01-275-3321	21	16	5320-00-956-7355	49	13
5310-01-275-3321	31		5320-00-956-7355	49	20
5310-01-275-3322	25	4	5320-00-956-7355	49	52
5310-01-275-3323	30	2	5320-00-957-3582	49	7
5310-01-275-3324	46	5	5320-00-957-3582	49	17
5310-01-275-3325	46	19	5320-00-957-3582	49	29
5310-01-275-3326	46	20	5320-00-957-3582	49	34
5310-01-275-3327	20	23	5320-00-957-3582	49	86
5310-01-275-7786	13	9	5320-00-971-7871	37	71
5310-01-275-7787	15	2	5320-00-971-7871	43	55
5310-01-276-1649	20	13	5320-01-004-0238	43	71
5310-01-276-1653	13	10	5320-01-049-8263	4	42
5310-01-276-1660	13	21	5320-01-049-8263	5	23
5310-01-276-1660	15	7	5320-01-049-8263	11	7
5310-01-276-1660	16	7	5320-01-049-8263	37	37
5310-01-276-1660	18	12	5325-00-099-8827	35	7
5310-01-276-1660	21	2	5325-00-099-8827	40	15
5310-01-276-1660	21	18	5325-00-099-8827	42	20
5310-01-276-3342	11	2	5325-00-174-9332	32	6
5310-01-276-8608	16	5	5325-00-174-9332	33	6
5310-01-277-7334	9	7	5325-00-185-0003	32	7
5310-01-277-7335	20	24	5325-00-185-0003	33	7
5310-01-278-8506	10	10	5325-00-185-0003	37	60
5310-01-282-5692	22	2	5325-00-185-0003	43	50
5310-01-282-7601	21	15	5325-00-432-9899	5	4
5310-01-301-1982	36	7	5325-00-432-9899	35	5
5310-01-301-1982	37	73	5325-00-432-9899	40	14
5310-01-301-1982	43	74	5325-00-432-9899	42	19
5310-01-301-1982	47	10	5325-00-449-2967	5	14
5315-00-449-2945	5	3	5325-00-449-2967	37	12

Section IV. NATIONAL STOCK NUMBER AND REFERENCE NUMBER INDEX

Table 3. National Stock Number to Figure and Item Number Index - Continued

NSN	FIGURE NO..	ITEM NO.	NSN	FIGURE NO.	ITEM NO.
5325-00-449-2967	43	6	5340-00-702-2848	40	12
5325-00-754-2187	37	59	5340-00-702-2848	42	12
5325-00-754-2187	43	49	5340-00-702-2848	43	47
5330-00-220-2631	7	8	5340-00-724-7038	8	11
5330-00-684-7851	6	26	5340-00-764-7051	37	46
5330-00-763-9322	7	3	5340-00-930-3386	32	12
5330-01-212-3361	6	39	5340-00-930-3386	33	12
5330-01-229-8077	19	5	5340-00-984-8540	11	11
5330-01-275-1956	30	4	5340-01-053-7130	35	4
5330-01-275-1956	31		5340-01-053-7130	37	11
5330-01-275-1961	30	8	5340-01-053-7130	40	4
5330-01-275-1961	31		5340-01-053-7130	42	4
5330-01-275-3338	25	1	5340-01-053-7130	43	5
5330-01-275-3338	31		5340-01-215-2927	17	17
5330-01-275-3357	5	5	5340-01-223-8726	17	12
5330-01-275-3358	21	4	5340-01-233-8305	37	44
5330-01-275-3358	31		5340-01-233-8305	43	35
5330-01-275-3359	21	20	5340-01-275-3401	13	20
5330-01-275-3359	31		5340-01-275-3401	15	9
5330-01-275-3360	22	5	5340-01-275-3404	6	15
5330-01-275-3360	31		5340-01-275-3477	47	21
5330-01-275-3361	22	11	5340-01-275-3526	13	18
5330-01-275-3361	31		5340-01-275-3527	5	16
5330-01-275-3368	6	30	5340-01-275-3528	20	36
5330-01-275-5013	27	8	5340-01-275-3529	8	9
5330-01-275-5013	31		5340-01-275-3534	37	42
5330-01-275-5014	27	9	5340-01-275-3534	43	31
5330-01-275-5014	31		5340-01-275-6025	6	14
5330-01-275-6832	31	11	5340-01-276-1795	24	5
5330-01-275-6832	31		5340-01-276-5908	25	7
5330-01-275-6838	27	12	5340-01-276-5912	21	10
5330-01-275-6838	31		5340-01-277-1185	25	8
5330-01-276-0897	18	9	5340-01-277-3363	47	9
5330-01-276-0897	31		5340-01-281-5270	37	68
5330-01-276-1681	31	27	5340-01-281-5270	43	52
5330-01-276-1681	31		5340-01-305-3401	6	24
5330-01-276-2290	25	10	5340-01-305-3406	22	4
5330-01-276-2290	31		5340-01-305-3414	46	3
5330-01-276-6707	10	14	5340-01-329-3933	19	4
5330-01-276-7501	20	35	5355-00-899-9014	35	21
5330-01-276-7501	31		5355-00-899-9014	40	25
5330-01-278-9475	10	11	5355-00-899-9014	42	21
5330-01-280-9392	9	9	5355-00-899-9014	42	25
5330-01-283-2407	10	9	5355-01-282-1863	13	17
5330-01-283-4297	12		5360-00-200-9691	7	10
5330-01-283-4297	31		5360-01-094-2717	7	9
5330-01-287-0922	9	12	5360-01-227-3195	17	20
5340-00-050-2740	37	25	5360-01-227-6302	17	8
5340-00-050-2740	47	3	5360-01-275-3506	24	3
5340-00-088-1254	4	40	5360-01-275-3513	20	28
5340-00-091-3790	35	10	5360-01-275-3514	20	29
5340-00-091-3790	37	3	5360-01-275-3516	20	15
5340-00-682-1617	52	4	5360-01-277-1192	20	30

Section IV. NATIONAL STOCK NUMBER AND REFERENCE NUMBER INDEX

Table 3. National Stock Number to Figure and Item Number Index - Continued

NSN	FIGURE NO.	ITEM NO.	NSN	FIGURE NO.	ITEM NO.
5365-00-804-7653	32	28	5935-00-115-2306	14	5
5365-00-804-7653	33	28	5935-00-115-2306	27	3
5365-01-173-3442	20	34	5934-00-115-2307	14	4
5365-01-274-9748	20	33	5935-00-295-6403	4	27
5365-01-275-6844	30	9	5935-00-564-5362	14	6
5365-01-275-6845	30	9	5935-00-567-0128	4	29
5365-01-275-6846	21	22	5935-00-567-0128	4	33
5365-01-275-6847	21	23	5935-00-815-1541	50	2
5365-01-275-6848	21	24	5935-01-012-3080	40	20
5365-01-275-6873	20	16	DELETED		
5365-01-275-7816	18	15	5935-01-070-3681	14	2
5365-01-275-7817	18	15	5935-01-175-0255	14	3
5365-01-275-7818	18	15	5935-01-219-4205	38	3
5365-01-275-7819	18	15	5935-01-219-4205	44	40
5365-01-277-4616	8	10	5935-01-280-1177	37	70
5365-01-281-1124	28	1	5935-01-280-1177	43	54
5365-01-281-1124	28		5940-00-113-8183	14	9
5365-01-282-2825	20	7	5940-00-113-8191	4	6
5365-01-282-2825	31		5940-00-113-8191	4	13
5640-01-287-2048	8	6	5940-00-113-8191	4	31
5905-00-139-1989	37	29	5940-00-113-8191	4	34
5905-00-139-1989	43	61	5940-00-113-9831	4	15
5905-00-643-5129	35	24	5940-00-115-2677	37	66
5905-00-643-5129	40	28	5940-00-115-2677	43	21
5905-00-643-5129	42	23	5940-00-115-5008	37	32
5905-00-883-8431	37	45	5940-00-115-5008	37	35
5905-00-883-8431	43	39	5940-00-143-4774	38	8
5910-01-279-0003	43	65	5940-00-143-4774	44	10
5910-01-280-0754	43	66	5940-00-143-4775	44	9
5910-01-280-0754	43	68	5940-00-143-4777	38	4
5910-01-280-0754	43	70	5940-00-143-4777	44	5
5910-01-283-6881	43	69	5940-00-143-4780	38	7
5910-01-283-9069	43	67	5940.00-143-4780	44	8
5920-00-284-9220	51	8	5940-00-143-4794	14	10
5920-00-892-9311	40	30	5940-00-143-4794	38	6
5920-00-892-9311	51	9	5940-00-143-4794	44	7
5920-01-113-2900	40	31	5940-00-204-8990	38	9
5925-00-682-0742	42	17	5940-00-230-0515	38	5
5925-00-686-3298	35	20	5940-00-230-0515	44	6
5925-00-686-3298	40	21	5940-00-283-5280	38	10
5925-00-686-3298	42	15	5940-00-283-5280	44	12
5925-00-961-1202	37	79	5940-00-477-9967	43	78
5925-00-966-5836	40	23	5940-00-549-6581	4	7
5925-00-984-4324	40	22	5940-00-549-6583	4	16
5925-00-984-4324	42	16	5940-00-682-2445	14	8
5930-00-259-4646	40	29	5940-00-738-6272	4	1
5930-00-259-4646	42	22	5940-00-788-1586	14	7
5930-00-538-5508	40	27	5940-00-825-3699	50	5
5930-00-538-5508	42	24	5940-00-825-3699	51	4
5930-00-659-2672	43	16	5940-00-836-0360	44	11
5930-01-055-9251	35	23	5940-00-952-2827	47	12
5930-01-055-9251	40	26	5940-00-958-0349	43	77
5930-01-280-5443	6	21	5940-00-958-1214	37	18
5940-00-983-6048	43	63			

Section IV. NATIONAL STOCK NUMBER AND REFERENCE NUMBER INDEX

Table 3. National Stock Number to Figure and Item Number Index - Continued

NSN	FIGURE NO..	ITEM NO.	NSN	FIGURE NO.	ITEM NO.
5940-00-983-6049	37	49	6115-01-289-1081	32	1
5940-00-983-6089	37	50	6115-01-292-6970	32	16
5940-00-983-6089	43	44	6115-01-295-8302	33	1
5940-00-983-6101	43	41	6115-01-298-1566	33	16
5940-00-983-6114	35	15	6125-00-659-2935	37	54
5940-01-054-6954	4	3	6140-00-059-3528	4	22
5940-01-112-9746	50	10	6145-00-189-6695	5	20
5940-01-115-5008	37	35	6145-00-578-6597	37	33
5940-01-126-3973	51	5	6145-00-578-6597	37	36
5945-00-435-1833	37	7	6145-00-578-7513	44	13
5945-00-435-1833	43	27	6145-00-578-7514	14	12
5945-00-686-6877	37	26	6145-00-578-7514	38	11
5945-00-686-6877	43	62	6145-00-578-7514	44	14
5945-01-213-9233	17	13	6145-00-655-2562	14	11
5945-01-280-5477	20	10	6145-00-655-2562	38	12
5950-01-054-4127	44	16	6145-00-655-2562	44	15
5961-00-712-5578	45	4	6145-01-047-0530	4	4
5961-00-724-5970	32	30	6145-01-047-0530	4	14
5961-00-724-5970	33	30	6145-01-047-0530	4	30
5970-00-787-2321	50	12	6145-01-047-0530	4	35
5970-00-787-2325	50	8	6150-00-519-2714	37	51
5970-00-787-2325	51	3	6150-00-519-2714	43	43
5970-00-914-3117	4	5	6150-00-632-7234	43	64
5970-00-929-5627	37	21	6150-00-949-0604	33	9
5940-00-954-1622	50	7	6150-01-051-0145	46	11
5975-00-074-2072	37	30	6150-01-274-5935	21	7
5975-00-074-2072	44	2	6150-01-280-0453	4	12
5975-00-074-2072	50	4	6150-01-286-9552	4	2
5975-00-074-2072	51	6	6625-00-054-2038	40	8
5975-00-111-3208	38	2	6625-00-055-9760	37	57
5975-00-111-3208	44	3	6625-00-065-5258	40	9
5975-00-136-9005	37	80	6625-00-065-5258	42	9
5975-00-727-5153	14	13	6625-00-065-5301	40	7
5975-00-727-5153	18	2	6625-00-065-5301	42	7
5975-00-878-3791	5	15	6625-00-065-8554	42	8
5975-00-879-7234	40	18	6625-00-068-0562	35	25
5975-00-924-9927	5	22	6625-01-157-9516	35	18
5975-00-944-1499	14	14	6645-00-089-8842	35	27
5977-01-224-2917	17	19	66454-0089-8842	40	10
5998-01-281-0071	37	5	6645-00-089-8842	42	10
5998-01-281-0071	43	25	6680-00-984-4745	35	26
5999-00-186-3912	5	21	6680-01-276-2683	19	1
6110-00-764-7621	45	1	6685-00-179-8632	9	14
6115-00-659-2786	43	37	9320-00-X88-3202	BULK	
6115-00-758-9239	32	9	9320-00-855-3399	BULK	
6115-00-859-2410	32	26	9320-00-641-3010	BULK	
6115-00-859-2410	33	26	9390-01-287-8896	5	2
6115-00-940-0175	43	36	9905-00-477-4131	37	81
6115-00-997-9769	32	9	9905-00-477-4137	47	14
6115-01-150-0367	3	1	9905-01-038-7439	32	14
6115-01-150-4140	1	1	9905-01-066-3081	32	14
6115-01-151-8126	2	1	9905-01-066-3081	33	14
6115-01-271-1584	48	1	9905-01-120-8728	5	24
6115-01-283-0414	33	27			

Table 4. Reference Number To Figure and Item Number Index

REFERENCE NUMBER FSCM NO.	FIGURE NO.	ITEM REFERENCE FIGURE ITEM NUMBER FSCM NO.				NO.	
AD42S	07707	49	58 GRADE M BLACK 81349		49	68	
AD43S	07707	49	23	GRADE M BLACK 81349	49	83	
ANSI B18.13	80204	36	3 GRADE M BLACK 81349 49 84				
ANSI B18.13	80204	37	1 GRADE M BLACK 81349 49 85				
ANSI B18.13	80204	40	24	GRADE M BLACK 81349 BULK			
ANSI B18.13	80204	42	13	GRADE M BLACK 81349 BULK			
ANSI B18.13	80204	43	46	GRB58 73616	5 22		
AN931A16-22	88044	49	62	J-C30AVA06CJ1	81348	5 20	
AN960-616L	88044	36	6 /6AVAB0				
B18C00590E	10983	43	12	JAN1N1204A	81349 32 30		
B18231A08025N	80204	31	1	JAN1N1204A	81349	33	30
CS4310-SV-0728	16236	9 11 JHP112-53 02032		37	18		
C0102114700	15434	6 39 1514	81343	6 33			
C0114031800	44940	28		KGDX2020	14655	50	13
C0718103700	15434	13	22	LF3525	33457 31 22		
C0718104800	15434	21	1	MEP-0168	30554	1	1
C07400100800	15434	30	2	MEP-021B	30554	2	1
C0740100400	15434	11	10	MEP-026B	30554	3	1
C0740100600	15434	13	21	MIL-B-5423-14	81349 40 22		
C0740100600	15434	15	7 MIL-C-5756		81349	4	4
C0750100600	15434	15	6 MIL-C-5756		81349	4	14
C0800205700	15434	15	17	MIL-C-5756 81349	4	30	
DG3M09F-S1-RPC	82168	43	77	MIL-C-5756	81349	4 35	
D15751	07860	32	9 MIL-C-5809		81349 40 21		
FA1796EEE-0111	01276	10	2 MIL-F-5591 81349		5	4	
FB9876-01-0070	01276	6 36 MIL-F-5591		81349	35	5	
FHN26G 1	81349	40	30	MIL-F-5591	81349 40 14		
FHN26G1	81349	51	9 MIL-F-5591		81349 42 19		
FIT2211-8 YELLOW	92194	50	8 MIL-S-19500-370		81349 45	4	
FIT2211-8 YELLOW	92194	51	3	MS 51957-83	96906	49	54
FIT2213-32	92194	50	12	MS14314-2X	96906 15 16		
YELLOW				MS 15795-409	96906	37	75
FIT2213/16 BLACK 92194	50	7	MS15795409		96906	43	76
FL3650EEE-0187	01276	6	3	MS15795-417 96906	37	17	
FL3650EEE-0307 01276	6 17	MS 15795-808 96906 49 3					
FL5114EEE-0091	01276	10	1	MS15795-808 96906	49	45	
FL5958EE-0263	01276	6 16	MS15795-808 96906 49 75				
FM09A250V12A	81349	40	31	MS15795-808	96906	52	3
FS0216B 122-1	15277	5 15	MS15795-810 96906 49 56				
GD1033G	74465	20	27	MS15795-841	96906	49	48
GRADE M BLACK	81349	49	5	MS 15795-841	96906 49 78		
GRADE M BLACK	81349	49	6	MS15795-910	96906 36	9	
GRADE M BLACK	81349	49	15	MS16624-1078	96906 32 28		
GRADE M BLACK	81349	49	16	MS16624-1078	96906 33 28		
GRADE M BLACK	81349	49	28	MS17143-10 96906	50	5	
GRADE M BLACK	81349	49	31	MS17143-10 96906 51	4		
GRADE M BLACK	81349	49	32	MS17143-15 96906 44 11			
GRADE M BLACK 6 1349	49	39	MS18154-60 96906 46		9		
GRADE M BLACK	81349	49	64	MS20426B4-6	96906	5 13	
GRADE M BLACK	81349	49	65	MS20426B4-6	96906	37	10
GRADE M BLACK	81349	49	66	MS204Z6B4-6	96906	43	4
GRADE M BLACK t 1349	49	67	MS20659-144		96906 37 66		

Table 4. Reference Number To Figure and Item Number Index

REFERENCE BER FSCM NO.	FIGURE NO.	ITEM REFERENCE FIGURE ITEM NUMBER FSCM NO.				NO.	NUM-
MS20659-144	96906	43	21	MS25244-7 1/2 96906	42	15	
MS21044C08	96906	49	47	MS25440-6 96906	46	2	
MS21044-C08	96906	49	77	MS25440-6	96906	46	10
MS21044-N3	96906	49	43	MS27"130S25	96906	52	17
MS21318-14	96906	32	13	MS27130-525	96906 52 21		
MS21318-14	96906	33	13	MS27130-534	96906 52 15		
MS21318-14	96906	47	13	MS27142-3	96906 14	5	
MS21333-102	96906	11	11	MS27142-3	96906 27	3	
MS21333-104	96906	4 40	MS27144-2	96906 14	4		
MS21333-128	96906	40	12	MS27183-10 96906	4 37		
MS21333-128	96906	42	12	MS27183-10 96906	5 10		
MS21333-128	96906	43	47	MS27183-10 96906	10	4	
MS21333-69	96906	37	46	MS27183-10 96906	46	13	
MS21333-72	96906	35	10	MS27183-10 96906	47	20	
MS21333-72	96906	37	3	MS27183-12 96906	4 19		
MS21333-75	96906	37	25	MS27183-12 96906	5 19		
MS21333-75	96906	47	3	MS27183-12 96906	5 27		
MS21333-76	96906	8 11	MS27183-12 96906		6 13		
MS21919DG12	96906	52	4	MS27183-12	96906	9	3
MS24166-D1	96906	37	26	MS27183-12 96906 31	3		
MS24166-D1	96906	43	62	MS27183-12	96906	37	41
MS24665-134	96906	11	1	MS27183-12 96906	43	30	
MS25002-1	96906	40	27	MS27183-12 96906	47	7	
MS25002-1	96906	42	24	MS27183-15 96906	32	19	
MS2S002-2 96906	40	29	MS27183-15	96906	33	19	
MS25002-2	96906	42	22	MS3367-1-9 96906	37	30	
MS25036-106	96906	38	10	MS3367-1-9 96906	44	2	
MS25036-106	96906	44	12	MS3367-1-9 96906	50	4	
MS25036-108	96906	38	7	MS3367-1-9 96906	51	6	
MS25036-108	96906	44	8	MS3367-4-9 96906	14	13	
MS25036- 11	96906	38	9	MS3367-4-9 96906	18	2	
MS25036-112	96906	14	10	MS3367-5-9 96906	38	2	
MS25036-112	96906	38	6	MS3367-5-9 96906	44	3	
MS25036-112	96906	44	7	MS3368-1-9A	96906	14	14
MS25036-113	96906	14	9	MS3450W32-7P	96906 38	3	
MS25036-127	96906	4	6	MS345OW32-7P	96906	44	4
MS25036-127	96906	4 13	MS3456W14S5S 96906		14	6	
MS25036-127	96906	4 31	MS3456W32-7S 96906 14		2		
MS25036127	96906	4 34	MS3476W8-4S	96906	50	2	
MS25036-128	96906	4 15	MS35190-254	96906 37 20			
MS25036-129	96906	37	32	MS35190-254	96906 43 73		
MS25036-129	96906	37	35	MS35206-243	96906 32 10		
MS25036-153	96906	38	8	MS35206-243	96906	33	10
MS25036-153	96906	44	10	MS35309-306	96906 36 10		
MS25036-154	96906	38	5	MS35333-108	96906	36	8
MS25036-154	96906	44	6	MS35333-40 96906	32	25	
MS25036-156	96906	44	9	MS35333-42 96906	36	5	
MS25036-157	96906	38	4	MS35335-35 96906	46	8	
MS25036-157	96906	44	5	MS35335-60 96906	52	7	
MS25036-158	96906	14	8	MS35535-89 96906	47	11	
MS25244-7 1/2	96906	35	20	MS35338-101	96906 43 75		
MS25244-7 1/2	96906	40	21	MS35338-120	96906 36 16		

Table 4. Reference Number To Figure and Item Number Index

REFERENCE NUMBER FSCM NO.	**FIGURE NO.**	**ITEM REFERENCE FIGURE ITEM NUMBER FSCM NO.**				**NO.**	
MS35338-120	96906	37	74	MSS1861-24 96906	52	5	
MS35338-124	96906	37	16	MS51958-63 96906	52	1	
MS35338-138	96906	49	2	MS51958-66 96906	49	1	
MS35338-138	96906	52	2	MSS1958-66	96906	49	44
MS35338-138	96906	52	19	MS51958-66 96906	52	18	
MS35338-139	96906	49	55	MS51967-2	96906 36 15		
MS35338-41	96906	45	3	MS51967-5	96906	4 11	
MS35338-42	96906	32	11	MS51967-6	96906	4	17
MS35338-42	96906	33	11	MS51967-6	96906	5	25
MS35338-44	96906	32	3	MS51967-6	96906	37	39
MS35338-44	96906	33	3	MS51967-6	96906	43	28
MS35338-44	96906	33	25	MS51967-6	96906 47	5	
MS35338-45	96906	4 18	MS51967-8	96906	4	9	
MS35338-45	96906	5 18	MS51967-9	96906 46	7		
MS35338-45	96906	5 26	MS51967-9	96906 46 17			
MS35338-45	96906	6 12	MS51968-6	96906	6	11	
MS35338-45	96906	9	2	MSS1968-8 96906	36	4	
MS35338-45	96906	15	3	MS5206-231 96906	45	2	
MS35338-45	96906	31	2	MS75004-1	96906	4	7
MS35338-45	96906	32	22	MS75004-2	96906	4	16
MS35338-45	96906	33	22	MS75047-1	96906	4	22
MS35338-45	96906	37	40	MS75058-1	96906	4	27
MS35338-45	96906	43	29	MS87006-13 96906	37	65	
MS35338-45	96906	47	6	MS87006-13 96906	43	20	
MS35338-46	96906	46	18	MS87006-3	96906	6	8
MS35338-47	96906	32	18	MS90725-3	96906 32	5	
MS35338-47	96906	33	18	MS90725-3	96906 33	5	
MS35356-35	96906	4 10	MS90725-31 96906		4 23		
MS35356-67	96906	4	8	MS90725-31 96906	37	43	
MS35489-123	96906	37	59	MS90725-31	96906	43	32
MS35489-123	96906	43	49	MS90725-6	96906 32 24		
MS35489-48	96906	32	6	MS90725-6	96906	33	24
MS35489-48	96906	33	6	MS90725-69 96906	46	1	
MS35643-1	96906	6 26	MS90728-13 96906		32	2	
MS35645-2	96906	6 27	MS90728-13 96906		33	2	
MS35649-2255N	96906	36	7	MS90728-32	96906	5 17	
MS35649-2255N	96906	37	73	MS90728-32	96906	9	1
MS35649-2255N	96906	43	74	MS90728-32	96906	32	21
MS35649-2255N	96906	47	10	MS90728-32	96906	33	21
MS35650-302	96906	20	20	MS90728-32 96906	47	8	
MS35650-3252	96906	20	31 MS90728-37 96906		5 28		
MS35650-3255	96906	20	38	MS91528-2K4B 96906	35	21	
MS35691-36	96906	37	15	MS91528-2K48 96906	40	25	
MS35823-6C	96906	35	4	MS91528-2K48	96906	42	21
MS35823-6C	96906	37	11	M16878/4BGBO	81349	50	9
MS35823-6C	96906	40	4 M16878/4BGBO		81349	51	2
MS35823-6C	96906	42	4 M16878/4BGB2		81349	50	14
MS35823-6C	96906	43	5 M16878/4GBG9 81349 50		6		
MS35842-11	96906	22	12	M22-03-00191FD 81349		35	24
MS35842-11	96906	53	2 M22-03-00191FD 81349		40	28	
MS35842-12	96906	52	9 M22-03-00191FD 81349		42	23 MS35842-13	
96906 47 16 M23053	5-103-4 92194	BULK MS35842-14		96906	9	4 M23053	
5-104-4 92194	BULK MS51504A8-4		96906	31	26		

Table 4. Reference Number To Figure and Item Number Index

REFERENCE BER FSCM NO.	FIGURE NO.	ITEM REFERENCE FIGURE ITEM NUMBER FSCM NO.				NO.	NUM-
M23053	5-105-0 92194 BULK		P-15121-78	45722	5	11	
M23053/5- 109-2	81349	4	5	P-15121-78 45722	6	2	
M23071/1-001	81349	49	50 P-15121-78	45722	6	38	
M23071/1-001	81349	49	80	P-15121-78 45722	8	3	
M24243/1-D402	81349	43	71	P-15121-78	45722	10	5
M24243/1-D403	81349	5	7	P-15121-78	45722	11	5
M24243/1-D403	81349	35	3	P-15121-78	45722	43	51
M24243/1-D403	81349	37	13	P-15121-78	45722	47	19
M24243/1-D403	81349	40	3	P-15121-79	45722	46	14
M24243/1-D403	81349	42	3	P-15121-82	45722	35	12
M24243/1-D403	81349	43	7	P-15121-82	45722	43	58
M24243/1-D405	81349	37	71	PR20-0112A4-1 82121	43	16	
M24243/1-D405	81349	43	55	P15121-17 45722	35	19	
M2423/6-A605H	81349	49	10	P15121-17	45722	42	14
M2423/6-A605H	81349	49	13	P15121-18	45722	37	52
M24243/6-A402H	81349	49	34	P15121-18	45722	43	23
M24243/6-A402H	81349	49	86	P15121-2	45722	37	6
M24243/6-A403H	81349	49	7 P15121-2 45722		37	28	
M24243/6-A403H	81349	49	17	P15121-2	45722	43	26
M24243/6-A403H	81349	49	20	P15121-20	45722	37	2
M24243/6-A403H	81349	49	29	P15121-20	45722	40	17
M24243/6-A403H	81349	49	40	P15121-20	45722	43	14
M24243/6-A403H	81349	49	52	P15121-21	45722	37	4
M24243/6-A403H	81349	49	69	P15121-21	45722	43	24
M24243/6-A604H	81349	49	11	P15121-22	45722	43	40
M24243/6-A604H	81349	49	13	P15121-3	45722	43	60
M3971/1-5	81349	35	37	P15121-33	45722	37	55
M3971/1-5	81349	40	10	P15121-33	45722	43	10
M5086/2-1-9	81349	37	33	P15121-34 45722 36 12			
M5086/2-1-9	81349	37	36	P15121-35 45722 37 22			
M5086/2-10-9	81349	44	13	P15121-35	45722 43 38		
M5086/2-12-9	81349	14	12	P15121-37	45722 35 14		
M5086/2-12-9	81349	38	11	P15121-37	45722 37 56		
M5086/2-12-9	81349	44	14	P15121-37	45722 43 42		
M5086/3-16-9	81349	14	11	P15121-48	45722 35	9	
M5086/3-16-9	81349	38	12	P15121-48	45722 37 23		
M5086/3-16-9	81349	44	15	P15121-48	45722 43	2	
M5423-14-07	81349	40	22	P15121-5	45722 35 17		
M5423-14-07	81349	42	16	P15121-5	45722 40	6	
M55164/28A-TBJB	81349	37	51	P15121-5	45722 42	6	
M55164/28A-TBJB	81349	43	43	P15121-50	45722 47	2	
M5757/23-003	81349	37	7 P15121-64	45722	8	8	
M5757/23-003	81349	43	27	Q01444	76700	9	7
M6000-E-00108	96906	53	3 Q04190	76700	9	12	
M8504952-1-8W	81349	50	3 Q106DL10399		44940	12	1
NAS 132903-80	80205	52	16	Q33639	76700	9	8
ONA-1005 16632	52	10	Q59400	76700	9	10	
P-15121-17	45722	40	19	Q66993	76700	9	9
P- 15121-50	45722	6 28 RA25177 59730 51		5			
P-15121-50	45722	37	24	RBX00-2280	18265	9 14	
P-15121-50	45722	43	34	RER75F4R02R	81349 37 29		
P-15121-78	45722	4 39 RER75F4R02R		81349 43 61			

Table 4. Reference Number To Figure and Item Number Index

REFERENCE FSCM NO.	FIGURE NO.	ITEM REFERENCE FIGURE ITEM NUMBER FSCM NO.				NO. NUMBER	
RE80G1R00	81349	37	45	1/8HHP-S	98441 31 12		
RE80G1R00	81349	43	39 100-1/16-B-3-ALUM 57137			5	2
RRC271	81348	6	9	1005169	18876	5	3
RV6304-2	53551	4 42	1005169	18876 35	6		
RV6304-2	53551	5 23	1005169	18876 40 13			
RV630-4-2	53551	11	7	1005169	18876 42 18		
RV630-4-2	53551	37	37	101-0473-00	44940	30	5
S-10445	55026	42	9 101-0473-00 44940		31	20	
S-10448	55026	40	8 101-0572 44940		31	18	
S-38230-G4	74159	47	12	101-0603	44940	31	9
SP2272FM	70411	6	7	101-0749	44940	31	21
SW-2	92194	50	16	101-0756	44940	31	5
SW-2	92194	BULK		101-0771	44940	31	5
S10450	55026	35	26	101005 0A569	10	14	
S6874N-2	51589	36	18	102-1147	44940	6	39
S6874P-2	51589	36	19	102-1259	44940 31 31		
TA03M72CR12	84971	8 10 102-1298 44940		6 37			
U1577	90788	54	4	102-1321-01	44940	8 12	
VS179-VL-4-2B	93061	20	8	102-1323 44940	31	25	
WC596/12-4	81348	40	20	102-1324 44940	31	24	
WW-P-471	81348	30	14	102-1344	44940	8	9
WW-P-471	81348	31	12	103-0763	44940	27	7
X-172	88786	49	26	103-0767	44940	27	11
X-172	88786	49	27	103-0782	44940	27	9
X-172	88786	BULK 103-0782		44940	31		
X121619	80072	6 10 103-0783 44940		27	12		
Z2258-9	76385	50	11	103-0783	44940	31	
000931 010355	64678	32	17	104-1564	44940	30	10
000931 010355	64678	33	17	104-1565-01	44940 30	9	
006-10196	74193	37	80	104-1565-02	44940	30	9
017536	50184	8 12	104-1631	44940 30 13			
030212	0A569	10	11	104-1635	44940 26	5	
031523	0A569	10	9 104-1636	44940	30	11	
0432 191 797	53867	21	13	104-1643	44940 26	6	
0433 117 120	53867	21	14	104-1646	44940 30 12		
099-2139-3	4	45		105-0587	44940	29	7
099-2316	44940	4 46 105-0599 44940		29	9		
099-2317	44094	4 43 105-0608 44940		29	5		
099-2318	44940	4 44 105-0609 44940		29	3		
099-2319-1	44940	4 45 10906258		19207 15 11			
099-2319-2	44940	4 45 10942521		19207	4	1	
099-2321	44940	11	8 10947657	19207	7 10		
099-2334	44940	43	9 110-2611	44940	24	1A	
099-2335	44940	37	14	110-2627 44940	24	5	
099-2343	44940	32	15	110-2725-01	44940	21	22
099-2343	44940	33	15	110-2725-02	44940	21	23
099-2381	44940	43	56	110-2725-03	44940	21	24
099-2382	44940	37	72	110-3176 44940	24	7	
099-2385	44940	43	8 110-3177	44940	24	9	
1-4X1-4CDB	98441	6	6	110-3178	44940 24	8	
1/4FG-B	98441	6 20	110-3179	44940 24	2		
1/8HHP-S	98441	30	14	110-3192	44940 24	6	

REFERENCE NUMBER	FSCM NO.	FIG. NO. NUMBER	ITEM NO.	REFERENCE NO.	FSCM	FIG.	ITEM
323-0329	44940 40 18 403-2359 44940			46	4		
323-0351	44940 40 20 403-2360 44940			46	16		
323-0682	44940 14	2	403-2361 44940		47	18	
323-0682	44940 14	6	403-2362 44940		5 16		
323-0705	44940 14	4	403-2363 44940		46	15	
323-1159	44940 14	3	403-2364 44940		5 29		
323-1326	44940	4	26	403-2365 44940	15	5	
326 896	00779	4	3 403-2371 44940		31	4	
332-0350	44940 14 10 403-2373 44940			15	4		
332-0941	44940 37 30 403-2374 44940			4 41			
332-0941	44940 44	2	403-2441 44940		13	20	
332-0942	44940 18	2	403-2441 44940		15	9	
332-1041	44940 37 50 403-2658 44940			47	21		
3321041	44940 43 44 405-3905 44940			5	5		
332-1043	44940 37 51 406-0304 44940			5 14			
332-1043	44940 43 43 406-0304 44940			37	12		
332-1046	44940 14	8	406-0304 44940		43	6	
332-1120	44940 44	9	406-0306 44940		5	3	
332-1290	44940 37 18 406-0306 44940			35	6		
332-1770	44940	5	15	406-0306 44940	40	13	
332-2073	44940 14	7	406-0306 44940		42	18	
332-2767	44940 37 70 406-0307 44940			5	4		
332-2767	44940 43 54 406-0307 44940			35	5		
333-0242	44940 21	8	406-0307 44940		40	14	
334-1283	44940 14 11 406-0307 44940			42	19		
334-1285	44940 14 12 406-0371 44940			35	7		
337-0098	44940 46 11 406-0371 44940			40	15		
337-2247	44940 37 76 406-0371 44940			42	20		
337-2248	44940 37 77 406-0581 44940			37	68		
337-2249	44940 37 78 406-0581 44940			43	52		
338-1896	44940 14	1	406-0583 44940		37	64	
338-1897	44940 44	1	406-0584 44940		4 21		
338-1899	44940 38	1	406-0643 44940		37	11	
345-029-27	70436 11	3	406-0643 44940		43	5	
356-0121	44940 37 47 406-0644 44940			42	4		
3598	07464	5	15	406-0853 44940	43	19	
363-0098	44940 36 23 416-0856 44940			4 12			
37TB7	81349 43 63 416-0857 44940			4	2		
37TB8	81349 37 49 416-0858 44940			4 25			
38TB12	81349 37 50 416-0859 44940			4 20			
38TB12	81349 43 44 416-0860 44940			4 24			
38TB3	81349 35 15 416-0868 44940			4	1		
39TB6	81349 43 41 41625		72850	7 11			
4-4 CTX-B	30780	6	23	420-0516 44940	48	1	
4-4 DTX-B	30780	6	4 420-0517 44940		48	2	
4-4 1303398	81343	6	10	420-0518 44940	48	3	
40128	72850	7	1 476285 72850		7 10		
402-0571	44940 37 42 4763096-1		28926	43	78		
402-0572	44940 47	9	479012 72850	7	4		

Table 4. Reference Number To Figure and Item Number Index

REFERENCE NUMBER	FSCM NO.	FIGURE NO.	ITEM NO.	REFERENCE NUMBER	FSCM NO.	FIGURE NO.	ITEM NO.
13211E6992-2	97403	42	8	147-4430	44940	31	
13212E8933	97403	43	37	147-0555	44940	21	10
13213E4091	97403	43	16	147-0714	44940	18	14
13213E4097	97403	32	26	147-4715-01	44940	21	13
13213E4097	97403	33	26	147-0721	44940	18	4
13213E4101	97403	33	9	147-0759	44940	25	7
13213E4107	97403	32	9	147-0760	44940	25	8
13213E4128	97403	44	16	147-0793	44940	21	15
13213E4168	97403	32	9	147-0793-01	44940	21	15
13213E4170	97403	33	31	147-0793-02	44940	21	15
13213E4222	97403	32	31	147-0793-03	44940	21	15
13213E4228	97403	32	12	147-079344	44940	21	15
13213E4228	97403	33	12	147-07934-5	44940	21	15
13213E4237	97403	37	54	147-0793-06	44940	21	15
13213E4264	97403	37	57	147-0793-07	44940	21	15
13214E6024	97403	32	29	147-40793-08	44940	21	15
13214E6024	97403	33	29	147-0793-09	44940	21	15
13214E6027	97403	32	14	147-0793-10	44940	21	15
13214E9580	97403	32	14	147-0793-11	44940	21	15
13214E9580	97403	33	14	147-0793-12	44940	21	15
1321SE1942	97403	43	78	147-0793-13	44940	21	15
13216E3329-1	97403	6	32	147-0793-14	44940	21	15
13216E3987	97403	37	21	147-0793-15	44940	21	15
13219E0883	97403	45	1	147-0793-16	44940	21	15
1344531	44940	26	2	147-0793-17	44940	21	15
1344533-01	44940	13	5	147-0793-18	44940	21	15
134-4543	44940	37	58	147-0793-19	44940	21	15
134-4543	44940	43	48	147-0793-20	44940	21	15
1344561	44940	26	3	147-0793-21	44940	21	15
134-4562	44940	13	14	147-0793-22	44940	21	1
134-4568	44940	8	7	147-0793-23	44940	21	15
134-4569	44940	8	6	147-0793-24	44940	21	15
134-4570	44940	8	5	147-0793-25	44940	21	15
134-4571	44940	8	4	147-0794	44940	21	14
134-4573-01	44940	13	4	149-1864	44940	18	10
134-4575-01	44940	13	15	149-2060	44940	18	9
134-4594	44940	13	7	149-2060	44940	31	
134-4597	44940	13	11	149-2105	44940	18	8
1344606-01	44940	13	3	149-2142	44940	18	
134-4684	44940	5	12	149-2142	44940	18	
134-4685	44940	5	6	149-2147	44940	10	16
134-4686-01	44940	5	1	150-0939	44940	20	19
134-4686-.02	44940	5	1	150-2013	44940	29	6
134-4687	44940	5	6	150-2044	44940	20	18
140-0961	44940	9	14	150-2045	44940	20	21
140-1990	44940	9	6	150-2084	44940	20	24
140-2039	44940	13	12	150-2109	44940	11	6
147-0405-01	44940	18	15	150-2126	44940	20	25
147-0405-02	44940	18	15	150-2127	44940	20	14
147-0405-03	44940	18	15	150-2129	44940	20	27
147-0405-04	44940	18	15	150-2130	44940	20	17
147-0430	44940	21	16	150-2132	44940	20	15

Table 4. Reference Number To Figure and Item Number Index

REFERENCE BER FSCM NO.	FIGURE NO.	ITEM REFERENCE FIGURE ITEM NUMBER FSCM NO.				NO.	NUM-
150-2133	44940	20	35	191-1609 44940	17	13	
150-2133	44940	31		191-1610	44940 16	8	
150-2135	44940	20	16 191-1612 44940		27	2	
150-2192	44940	20	28	191-1618 44940	17	1	
150-2228	44940	20	29	191-1647 44940	16	3	
150-2229	44940	20	30	191-1848 44940	17	3	
150-2230	44940	20	32	191-1859 44940	17	10	
15001	74400	35	27	191-1860	44940 17	4	
15001	74400	40	10	191-1861	44940 17 18		
15001	74400	42	10	191081	0A569 10 13		
151051	0A569	10	7 192001 0A569		10	10	
152-0260	44940	20	36	192012	0A569 10 12		
152-0262	44940	20	3 192050 0A569		10		
153022	0A569	10	8 192100 0A569		10	6	
154-2322	44940	21	4 2-430-101-035		53867 21 15		
154-2322	44940	31		2-430-101-037	53867	21	15
154-2342	44940	21	20	2-430-101-039	53867	21	15
154-2342	44940	31		2-430-101-041	53867 21 15		
154-2636	44940	21	3 2-430-101-043		53867 21 15		
154-2646	44940	21	19	2-430-101-045	53867	21	15
155-2088	44940	15	18	2-430-101-047	53867	21	15
155-2091	44940	15	12	2-430-101-049	53867	21	15
155-2098	44940	15	15	2-430-101-051	53867	21	15
155-2147	44940	15	14	2-430-101-053	53867	21	15
155-2163	44940	15	10	2430-101-055	53867	21	15
159-1135	44940	6 14 2430-101-057		53867	21	15	
159-1136	44940	6 25 2-430-101-059		53867 21 15			
159-1137	44940	6 24 2-430-101-061		53867	21	15	
159-1138	44940	6 15 2-430-101-063		53867	21	15	
159-1142	44940	6 22 2-430-101-065		53867	21	15	
159-1143	44940	6 18 2-430-101-067		53867	21	15	
159-1175	44940	6 29 2-430-101-069		53867	21	15	
159-1176	44940	6 30 2-430-101-071		53867	21	15	
160-1340	44940	13	17	2-430-101-073	53867	21	15
168-0184	44940	12		2-430-101-075	53867 21 15		
168-0184	44940	31		2-430-101-077	53867 21 15		
169VL-4-2	93061	20	9 2-430-101-079 53867		21	15	
19G	70040	21	8 2-430-101-081		53867	21	15
191-1432	44940	17	17	2-430-101-083	53867	21	15
191-1433	44940	17	20	2-520184-2 00779	50	10	
191-1436	44940	17	16	2AN44	45152	4	38
191-1437	44940	17	15	2N3442	81349	45	4
191-1439	44940	17	12	2P1279	11083	6	35
191-1440	44940	17	8 20-1085-2	94916	43	37	
191-1441	44940	17	9 200-0999	44940	32	1	
191-1442	44940	17	5 200-1000	44940	32	1	
191-1444	44940	17	7 200-1023	44940	33	1	
191-1446	44940	17	11	201-1555 44940	32	31	
191-1450	44940	17	2 201-1557	44940	32	16	
191-1605	44940	17	6 201-1673	44940	33	16	
191-1606	44940	17	14	201-1674 44940	33	31	
191-1608	44940	17	19	201-3417 44940	32	27	

Table 4. Reference Number To Figure and Item Number Index

REFERENCE NUMBER FSCM NO.	FIGURE NO.	ITEM REFERENCE FIGURE ITEM NUMBER FSCM NO.				NO.	
201-3418	44940	33	27	303737	32195 37 61		
205-0079	44940	32	20	303737	32195 43 17		
205-0079	44940	33	20	30397	14655 50 15		
205K1114	94990	32	7 305-0733	44940	36	1	
205K1114	94990	33	7 305-0771	44940	36	13	
205K1114	94990	37	60	305-0772 44940	36	20	
205K1114	94990	43	50	305-0773 44940	36	17	
2110-0312	28480	51	8 305-0774	44940	36	14	
22006	26405	43	41	305-0775	44940 36 22		
226-3078	44940	4 32 305-0776 44940		36	21		
226-3079	44940	4 28 307-2386 44940		20	10		
226-3310	44940	21	7 307-2387	44940	20	11	
226-3693	44940	37	34	308-0318 44940	35	23	
226-3695	44940	37	31	308-0318 44940	40	26	
231-0271	44940	30	7 312-0252	44940	37	53	
231-0272	44940	30	3 312-0260-02	44940	43	66	
232-2009	44940	33	26	312-0260-02	44940	43	68
232-2015	44940	32	12	312-0260-02	44940	43	70
232-2015	44940	33	12	319-0186 44940	40	2	
232-3228	44940	32	23	319-0187 44940	43	3	
232-3228	44940	33	23	319-0188 44940	42	2	
2322009	44940	32	26	319-0189	44940	35	2
234-0805	44940	32	4 319-0190	44940	37	9	
234-0805	44940	33	4 319-0192	44940	47	4	
234-0820	44940	32	8 319-0193	44940	43	15	
234-0820	44940	33	8 319-0194	44940	37	69	
25012-1	81860	47	9 319-0194	44940 43 53			
2860	70485	32	7	323-0682	44940 14	6	
2860	70485	33	7	323-1326	44940	4	26
2860	70485	37	60	326 896	00779	4	3
2860	70485	43	50	332-2767	44940 37 70		
29031	81860	37	42	332-2767	44940 43 54		
29031	81860	43	31	333-0242	44940 21	8	
294142	06853	4 38	3374-0098	44940 46 11			
295901-2	60119	35	7 337-2247	44940	37	76	
295901-2	60119	40	15	337-2248 44940	37	77	
295901-2	60119	42	20	337-2249 44940	37	78	
2965142	77060	14	7	338-1896	44940 14	1	
3/8X1/4FFB	98441	6 33 338-1897 44940		44	1		
300-2953	44940	37	5 338-1899	44940	38	1	
300-2953	44940	43	25	345-029-27 70436	11	3	
300-2962	44940	34	1 356-0121	44940	37	47	
300-2963-0!	44940	39	1	3598 07464	5 15		
300-2963-02	44940	41	1	363-0098	44940	36	23
301-0736	44940	35	22	37TB7 81349	43	63	
301-9273	44940	37	82	37TB8 81349	37	49	
301-9285	44940	43	33	38TB12	81349	37	50
301-9323	44940	35	1 38TB12 81349		43	44	
301-9324	44940	40	1 38TB3	81349	35	15	
301-9325	44940	42	1 39TB6F 81349		43	41	
301-9478	44940	37	62	4-4 130339B	81343	6 10	
301-9483	44940	43	18	4-4 140139B	81343	6 20	

Table 4. Reference Number To Figure and Item Number Index

REFERENCE BER FSCM NO.	FIGURE NO.	ITEM REFERENCE FIGURE ITEM NUMBER FSCM NO.				NO.	NUM-
4-4BTX-B	98441	6	4	479132	72850	7	9
4-4CBTXB	98441	6 23	479136	72850	7	3	
4-4070202BA	81343	6 23	479137	72850	7	6	
4-4070203BA	81343	6	4	479138	72850	7	7
40128	72850	7	1	479139	72850	7	8
402-0572	44940	47	9 479729	72850	7	5	
402-0573	44940	46	20	4917260-05	90536	52	13
403-2358	44940	46	3 5-4GBTX	98441	6	5	
403-2359	44940	46	4 501-040800-00		78189	35	16
403-2360	44940	46	16	501-040800-00	78189	37	27
403-2361	44940	47	18	501-040800-00	78189	40	5
403-2362	44940	5 16 501-040800-00		78189	42	5	
403-2363	44940	46	15	501-040800-00	78189	43	57
403-2364	44940	5 29 501-0504 44940		6 17			
403-2365	44940	15	5 501-0506	44940	6 16		
403-2371	44940	31	4 501-115	44940 47 17			
403-2373	44940	15	4 501-250800-00		78189	4 36	
403-2374	44940	4 41 501-250800-00		78189	5	9	
403-2441	44940	13	20	501-250800-00	78189	6	1
403-2441	44940	15	9 501-250800-00		78189	10	3
403-2658	44940	47	21	501-250800-00	78189	11	4
405-3905	44940	5	5	501-250800-00	78189	35	11
406-0371	44940	35	7 501-250800-00		78189	36	2
406-0371	44940	40	15	501-250800-00	78189	37	63
406-0371	44940	42	20	501-250800-00	78189	42	11
406-0581	44940	37	68	501-250800-00	78189	43	45
406-0581	44940	43	52	501-250800-00	78189	46	12
406-0583	44940	37	64	501-250800-00	78189	47	1
406-0584	44940	4 21 502-0153 44940		15	16		
406-0853	44940	43	19	502-0318 44940	6	7	
416-0856	44940	4 12 502-0951 44940		20	8		
416-0857	44940	4	2 502-0952	44940	20	5	
416-0858	44940	4 25 502-0953 44940		20	7		
416-0859	44940	4 20 502-0953 44940		31			
416-0860	44940	4 24 502-0977 44940		20	9		
41625	72850	7 11 502-0979 44940		31	32		
420-0516	44940	48	1 502-1015	44940	31	26	
420-0516	44940	54	1 502-1018	44940	18	5	
420-0517	44940	48	2 503-1055	44940	18	3	
420-0517	44940	54	2 503-1401	44940	15	13	
420-0518	44940	48	3 503-1440	44940	22	13	
420-0518	44940	54	3 503-1445-01	44940	18	1	
4217S8	23040	37	1 503-1464	44940	9	5	
42817S8	23040	36	3 503-1517-01	44940	8	1	
42817S8	23040	40	24	503-15174-01	44940	8	2
42817S8	23040	42	13	509-0142 44940	19	5	
42817S8	23040	43	46	509-0163 44940	30	8	
430-008	72741	21	9 509-0163	44940 31			
4431	13483	43	36	509-0166	44940 27	8	
476285	72850	7 10	509-0166	44940 31			
479012	72850	7	4	509-0204	44940	30	4
479130	72850	7	2	509-0204	44940	31	

Table 4. Reference Number To Figure and Item Number Index

REFERENCE NUMBER FSCM NO.	FIGURE NO.	ITEM REFERENCE FIGURE ITEM NUMBER FSCM NO.				NO.	
509-0211	44940	31	526-0313	44940	21	6	
509-0211	44940	31	11	526-1023 44940	46	5	
509-0221	44940	24	4 526-1024	44940	46	19	
509-0221	44940	31		526-2109	44940 15	2	
509-0228	44940	31		526-2133	44940 23	2	
509-0228	44940	31	27	6WGBTXS	98441	6 34	
509-0236	44940	10	9 6WLNS	30780	6 35		
509-0237	44940	10	11	600J	83330 43 64		
509-0238	44940	10	14	6202	81646 18 13		
510-0097	44940	32	29	6294493	77060 14	3	
510-0097	44940	33	29	68HB-8-4	93061 31 32		
510-0151	44940	31	13	69-539-2 30554	6	7	
511-061800-00	78189	37	48	69-561-1	30554 42	5	
511-061800-00	78189	40	16	69-561-1	30554 35 16		
511-061800-00	78189	43	13	69-561-1	30554 37 27		
511-081800-00	78189	35	13	69-561-1	30554 40	5	
511-081800-00	78189	36	11	69-561-1	30554 43 57		
511-081800-00	78189	37	19	69-561-2	30554 37 48		
511-081800-00	78189	43	11	69-561-2	30554 40 16		
511-101800-00	78189	35	8	69-561-2 30554	43	13	
511-101800-00	78189	37	8	69-561-3 30554	35	13	
511-101800-00	78189	47	22	69-561-3	30554 36 11		
511-101800-00	78189	40	11	69-561-3	30554 37 19		
511-101800-00	78189	43	1	69-561-3 30554	43	11	
515-1	44940	29	8 69-561-4 30554		35	8	
515-1	44940	30	16	69-561-4	30554	37	8
516-2003	44940	30	6 69-561-4 30554		43	1	
516-2003	44940	31	7 69-561-4 30554		47	22	
517-0217	44940	31	15	69-561-5 30554	4 36		
517-0218	44940	31	6 69-561-5 30554		5	9	
517-0219	44940	31	8 69-561-5 30554		6	1	
517-0230	44940	13	18	69-561-5 30554	10	3	
517-0235	44940	22	9 69-561-5 30554		11	4	
518-0034	44940	20	33	69-561-5 30554	35	11	
518-0207	44940	20	34	69-561-5 30554	36	2	
518-0279	44940	32	28	69-561-5 30554	37	63	
518-0279	44940	33	28	69-561-5 30554	40	11	
518-0399	44940	28		69-561-5	30554	42	11
518-0399	44940	28		69-561-5	30554	43	45
518-0399	44940	28		69-561-5	30554	46	12
518-0399	44940	28	1 69-561-5 30554		47	1	
52-132-1MG3	08602	37	79	69-662-17	30554	35	19
520-2202	44940	21	11	69-662-17	30554	40	19
520-2202	44940	21	17	69-662-17	30554	42	14
520-2206	44940	15	8 69-662-18	30554	37	52	
520-2408	44940	31	14	69-662-18	30554	43	23
5211	74545	40	18	69-662-2	30554	37	6
52305A18-6	84256	52	20	69-662-2	30554 37 28		
526-0016	44940	11	2 69-662-2 30554		43	26	
526-0018	44940	13	9 69-662-20	30554	37	2	
526-0174	44940	46	21	69-662-20	30554	40	17
526-0240	44940	46	6 69-662-20	30554	43	14	

Table 4. Reference Number To Figure and Item Number Index

REFERENCE BER FSCM NO.	FIGURE NO.	ITEM REFERENCE FIGURE ITEM NUMBER FSCM NO.				NO.	NUM-
69-662-21	30554	37	4 718-1021 44940		29	1	
69-662-21	30554	43	24	718-1022	44940	25	5
69-662-3	30554	43	60	718-1025	44940	27	5
69-662-33	30554	37	55	718-1027	44940	13	19
69-662-33	30554	43	10	718-1027	44940	27	4
69-662-34	30554	36	12	718-1037	44940	13	22
69-662-35	30554	37	22	718-1040	44940	16	6
69-662-35	30554	43	38	718-1046	44940	22	AI
69-662-37	30554	35	14	718-1048	44940	21	1
69-662-37	30554	37	56	718-1055	44940	30	1
69-662-37	30554	43	42	72-169-2MG6	74193 40 23		
69-662-48	30554	35	9 72-170-1MGL 74193 42 17				
69-662-48	30554	37	23	72-5006-1 30554	37	44	
69-662-48	30554	43	2 72-5006-1	30554	43	35	
69-662-5	30554	35	17	72-5006-2	30554	46	11
69-662-5	30554	40	6	72-5011	30554	35	23
69-662-5	30554	42	6	72-5011	30554	40	26
69-662-50	30554	6 28	72-5017	30554	11	4	
69-662-50	30554	37	24	72-5018	30554	4 42	
69-662-50	30554	43	34	72-5018	30554	5 23	
69-662-50	30554	47	2 72-5018	30554	11	7	
69-662-52	30554	43	40	72-5018	30554	37	37
69-662-64	30554	8	8	72-5029	30554	5 24	
69-662-78	30554	4 39	72-5101	30554	6 19		
69-662-78	30554	5 11	72-5217-1 30554		46	6	
69-662-78	30554	6	2	72-5319	30554	4	3
69-662-78	30554	6 38	720-1037	44940	18	11	
69-662-78	30554	8	3	720-1806 44940	16	4	
69-662-78	30554	10	5 725-1029 44940		20	1	
69-662-78	30554	11	5 725-1031 44940		20	26	
69-662-78	30554	43	51	740-1004	44940	11	10
69-662-78	30554	47	19	740-1004	44940	13	2
69-662-79	30554	46	14	740-1004	44940	16	2
69-662-82	30554	35	12	740-1004	44940	18	7
69-662-82	30554	43	58	740-1004	44940	20	2
69-668	30554	47	17	740-1004	44940 25	6	
69-695	30554	5	3	740-1004	44940 27	6	
69-695	30554	35	6 740-1004	44940 29	2		
69-695	30554	40	12	740-1004	44940 31 30		
69-695	30554	42	18	740-1006	44940 13 21		
69-766	30554	40	15	740-1006	44940 15	7	
69-766	30554	42	20	740-1006	44940 16	7	
69-788	30554	9 14 740-1006 44940 18 12					
695	79470	6 19 740-1006 44940		21	2		
70-801074	04655	5 21 740-1006 44940		21	18		
718-1018	44940	11	9 740-1008	44940	30	2	
718-1018	44940	13	1 740-1010	44940	16	5	
718-1018	44940	20	4 740-1802	44940	22	2	
718-1018	44940	31	29	75-5074	30554 47 15		
718-1020	44940	26	1 750-1004	44940	16	1	
718-1020	44940	27	10	75509-N 76700	9	6	
718-1021	44940	18	6 7598948	19200	35	5	

Table 4. Reference Number To Figure and Item Number Index- Continued

REFERENCE NUMBER	FSCM NO.	FIGURE NO.	ITEM NO.	REFERENCE NUMBER	FSCM NO.	FIGURE NO.	ITEM NO.
76902LA	02121	35	23	84-13027	30554	46	3
775-0035	44940	30	15	84-13028	30554	46	4
775-0073	44940	31	17	84-13030	30554	15	5
775-0075	44940	31	16	84-13032	30554	46	16
775-0076	44940	21	12	84-13033	30554	46	15
800-1005	44940	32	24	84-13034	30554	47	9
800-1005	44940	33	24	84-13037	30554	4	43
800-2059	44940	15	1	84-13038	30554	4	44
800-4074	44940	20	22	84-13038	30554	49	18
802-2505	44940	29	4	84-13041	30554	46	20
806-2004	44940	26	4	84-13042	30554	4	45
815-0333	44940	19	2	84-13042	30554	4	45
815-0599	44940	27	1	184-13042	30554	4	45
815-0627	44940	13	16	84-13043	30554	46	5
815-0633	44940	20	37	84-13044	30554	46	19
815-0638	44940	22	3	84-13048	30554	4	2
815-0672	44940	13	13	84-13049	30554	4	12
815-0673	44940	13	6	84-13050	30554	6	24
815-0674	44940	13	8	84-13051	30554	6	25
818-0076	44940	4	42	84-13054	30554	36	13
818-0076	44940	5	23	84-13055	30554	36	20
818-0076	44940	11	7	84-13056	30554	36	17
818-0076	44940	37	37	84-13057	30554	36	14
8198810	19200	37	67	84-13058	30554	36	22
8198810	19200	43	22	84-13059	30554	36	21
8201-02	03007	49	25	84-13060	30554	15	12
8201-02	03007	49	42	84-13061	30554	4	25
8201-02	03007	49	71	84-13062	30554	4	32
8203-02	03007	49	11	84-13063	30554	4	28
8203-02	03007	49	14	8413064	30554	4	26
8207-02	03007	49	24	84-13067	30554	36	1
8207-02	03007	49	41	84-13068-1	30554	36	19
8207-02	03007	49	70	84-13069	30554	37	64
8208-02	03007	49	21	84-13069	30554	43	19
8208-02	03007	49	53	84-13070	30554	35	1
84-10389	30554	41	1	84-13071	30554	37	53
84-13000	1DS87	6	31	84-13073	30554	6	18
84-13000	30554	6	31	84-13074	30554	11	3
84-13001	30554	6	21	84-13075	30554	4	20
84-13002	30554	6	22	84-13076	30554	4	24
84-13005	30554	6	14	84-13077	30554	31	4
84-13006	30554	6	15	84-13078	30554	4	21
84-13008	30554	6	3	84-13081	30554	11	6
84-13009	30554	6	17	84-13082	30554	11	8
84-13010	30554	10	1	84-13083	30554	5	16
84-13011	30554	6	16	84-13084	30554	4	41
84-13013	30554	47	18	84-13086	30554	37	47
84-13014	30554	4	46	84-13087	30554	34	1
84-13014	30554	49	8	84-13088	30554	40	1
84-13015	30554	37	42	84-13089-01	30554	39	1
84-13015	30554	43	31	84-13089-02	30554	41	1
84-13020	30554	5	29	84-13092	30554	15	4

Table 4. Reference Number To Figure and Item Number Index

REFERENCE BER FSCM NO.	FIGURE NO.	ITEM REFERENCE FIGURE ITEM NUMBER FSCM NO.				NO.	NUM-
84-13098	30554	6 36 84-13224 30554		5	5		
84-13109	30554	8	5	84-13225-01	30554	5	1
84-13111	30554	8	7	84-13225-02	30554	5	1
84-13117	30554	42	1 84-13278 30554		8	9	
84-13118	30554	14	1 84-13279 30554		47	4	
84-13120	30554	14	3 84-13281 30554		40	2	
84-13122	30554	32	20	84-13284 30554	43	15	
84-13122	30554	33	20	84-13285 30554	43	3	
84-13123	30554	32	4 84-13286 30554		6 29		
84-13123	30554	33	4 84-13287 30554		6 30		
84-13124	30554	33	16	84-13288 30554	42	2	
84-13125	30554	32	16	84-13289 30554	35	2	
84-13126	30554	32	8 84-13290 30554		9	6	
84-13126	30554	33	8 84-13291 30554		9	5	
84-13127	30554	32	23	84-13292 30554	37	9	
84-13127	30554	33	23	84-13293 30554	37	69	
84-13134	30554	33	1 84-13293 30554		43	53	
84-13135	30554	32	1 84-13294 30554		43	8	
84-13136	30554	32	1 84-13295 30554		43	56	
84-13139	30554	7	1	84-13296	30554	37	72
84-13142	30554	8	4	84-13301	30554	37	34
84-13144	30554	37	62	84-13303 30554	37	31	
84-13146	30554	37	14	84-13310-01	30554	43	69
84-13147	30554	43	9 84-13310-02	30554	43	67	
84-13148	30554	43	18	84-13310-03	30554	43	65
84-13154	30554	44	1 84-13311-02	30554	43	66	
84-13156	30554	38	1 84-13311-02	30554	43	68	
84-13162	30554	43	33	84-13311-02	30554	43	70
84-13168-2	30554	36	18	84-13317	30554	5	2
84-13171	30554	37	70	84-13320 30554	10	13	
84-13171	30554	43	54	84-13322 30554	10	12	
84-13176	30554	37	82	84-13324 30554	10		
84-13178	30554	37	5 84-13326 30554		10	8	
84-13178	30554	43	25	84-13327 30554	10	7	
84-13181	30554	37	58	84-13328 30554	9	7	
84-13181	30554	43	48	84-13329 30554	9	8	
84-13183	30554	35	22	84-13330 30554	9 10		
84-13183	30554	43	59	84-13331 30554	9 13		
84-13187	30554	36	23	84-13334 30554	10	2	
84-13190	30554	37	68	84-13335 30554	9	9	
84-13190	30554	43	52	84-13336 30554	9 11		
84-13195	30554	37	77	84-13337 30554	9 12		
84-13196	30554	37	76	84-13338 30554	8 10		
84-13197	30554	37	78	84-13339 30554	5	6	
84-13205	30554	6 37 84-13549 30554		49	61		
84-13208	30554	10	6 850-0055	44940 33 18			
84-13209	30554	12	1 850-1040	44940	32	3	
84-13210	30554	32	15	850-1040 44940	33	3	
84-13210	30544	33	15	850-1040 44940	33	25	
84-13218	30554	5 12 850-1045 44940		4 18			
84-13220	30554	5	6	850-1045	44940	5 18	
84-13221	30554	47	21	850-1045 44940	5 26		

Table 4. Reference Number To Figure and Item Number Index

REFERENCE NUMBER FSCM NO.	FIGURE NO.	ITEM REFERENCE FIGURE ITEM NUMBER FSCM NO.				NO.	
850-1045	44940	6 12 88-13555 30554		49	30		
850-1045	44940	9	2	88-13558	30554	49	59
850-1045	44940	15	3 88-13560 30554		48	1	
850-1045	44940	31	2 88-13561 30554		49	57	
850-1045	44940	32	22	88-13561 30554	49	73	
850-1045	44940	33	22	88-13562 30554	49	82	
850-1045	44940	37	40	88-13563 30554	49	33	
850-1045	44940	43	29	88-13564 30554	49	12	
850-1045	44940	47	6 88-13565 30554		49	4	
850-1055	44940	32	18	88-13566 30554	49	22	
850-2004	44940	20	23	88-13587-001	30554	52	6
860-2058	44940	21	9 88-13587-002	30554	49	74	
862-1001	44940	36	15	88-13596-001	30554	49	51
869-0006	44940	10	10	88-13596-001	30554	49	81
870-0131	44940	20	12	88-13597 30554	49	79	
870-0440	44940	13	10	88-13598-001	30554	49	76
870-1221	44940	35	13	88-13598-002	30554	49	46
870-1221	44940	36	11	88-13603 30554	49	35	
870-1221	44940	37	19	88-13605 30554	50	1	
870-1221	44940	43	11	88-13612 30554	52	11	
870-1232	44940	40	11	88-13613 30554	52	8	
870-2052	44940	20	13	88-13616 30554	49	36	
871-0018	44940	20	38	88-13617 30554	52	14	
88-13529	30554	51	7 915025-0066	65597	49	72	
88-13530-1	30554	52	12	982-92-2-22-0 60119 40 14			
88-13540	30554	51	1 982-92-2-220	60119	5	4	
88-13544	30554	53	1 98292-2-220	60119	42	19	
88-13546-4	30554	49	63	99836	60119	5	3
88-13546-5	30554	49	87	99836	60119 35	6	
88-13547-2	30554	49	37	99836	60119 40 13		
88-13547-3	30554	49	9	99836	60119	42	18
88-13547-5	30554	49	19	99947P130 60119	5 14		
88-13547-6	30554	49	60	99947P130 60119	37	12	
88-13548	30554	49	38	99947P130 60119	43	6	

Section V. REFERENCE DESIGNATOR INDEX

Table 5. Reference Designator Index

REFERENCE DESIGNATOR	FIGURE & ITEM NO.	REFERENCE NUMBER	REFERENCE DESIGNATOR	FIGURE & ITEM NO.	REFERENCE NUMBER

THIS SECTION NOT APPLICABLE.

DPSGAFB, AL (93000670) 1100